Kurt L. Loening (Ed.)

The Terminology of Biotechnology: A Multidisciplinary Problem

Proceedings of the 1989 International Chemical Congress of Pacific Basin Societies PACIFICHEM '89

With 44 Figures

Springer-Verlag
Berlin Heidelberg NewYork
London Paris Tokyo Hong Kong Barcelona

Dr. Kurt L. Loening
Managing Director
Topterm
North American Division
P.O Box 21213
Columbus, Ohio 43221
U.S.A

ISBN-13: 978-3-540-53052-7 e-ISBN-13: 978-3-642-76011-2
DOI: 10.1007/978-3-642-76011-2

Library of Congress Cataloging-in-Publication Data
International Chemical Congress of Pacific Basin Societies (1989)
The terminology of biotechnology: a multidisciplinary problem:
proceedings of the 1989 International Chemical Congress of Pacific Basin Societies.
PACIFICHEM '89 / Kurt Loening (editor).
Includes index.

1. Biotechnology--Terminology--Congresses. I. Loening, Kurt. II. Title.
TP248.16.I57 1989
660'.6'014--dc20 90-10301

Bookbinding: Lüderitz & Bauer GmbH, Berlin
2151/3020-543210 – Printed on acid-free paper

Preface

Biotechnology is such a magic word these days. Even though many people have defined it in different ways, it is clearly multidisciplinary in nature with contributions from many fields, such as biochemistry, molecular biology, genetics, immunology, microbiology, pharmacology, fermentation, and agriculture. Each of these subject areas brings with it its own special vocabulary and nomenclature standards. This results in considerable difficulties of communication between scientists of the different disciplines as well as between scientists and the general public.

The symposium was organized to address this problem by bringing together people from some of the major disciplines that contribute to biotechnology who are knowledgeable about nomenclature and terminology in their fields. As a result, the symposium identified some of the areas in biotechnology where proper nomenclature needs to be publicized. In addition, current terminology problems among the related, but diverse disciplines were brought out. Also, the role of the fields of terminology and computer science in overcoming these communication barriers was explored.

That the topic is of great interest to many people was demonstrated to me by the fact that 26 organizations provided financial support for the symposium, and I want to express here again my sincere gratitute to all of them. Special thanks are due to my co-organizers, Professor Akira Tsugita, Life Science Institute, Science University of Tokyo, and Dr. Joshua Rokach, Florida Institute of Technology, Melbourne, Florida. Finally, a special thank you to Joy Merritt, Chemical Abstracts Service, without whose help the whole undertaking could never have been completed so successfully.

Kurt L. Loening

Table of Contents

Activities of the IUPAC/IUB Joint Commission on Biochemical Nomenclature (JCBN) and the Nomenclature Committee of IUB (NC-IUB)

Kurt L. Loening

Chemical Abstracts Service®,
P.O. Box 3012, Columbus, Ohio 43210

Abstract: A brief history of biochemical nomenclature and the development of joint nomenclature efforts of the International Union of Pure and Applied Chemistry (IUPAC) and the International Union of Biochemistry (IUB) are presented. The organizational arrangement of cooperation between the two unions is described. Results of this setup are listed. Special mention is made of the biochemical compendium and the enzyme nomenclature book, and future projects are outlined. Cooperation with other nomenclature bodies that have overlapping interests is cited, such as with the International Union of Nutritional Sciences (IUNS), and with the International Federation of Clinical Chemistry (IFCC).

JOINT ENTERPRISE OF TWO UNIONS

One of the reasons that I put the topic of "Activities of the IUPAC/IUB Joint Commission on Biochemical Nomenclature (JCBN) and the Nomenclature Committee of IUB (NC-IUB)" first is that this is one of the longest ongoing or perhaps the only ongoing joint enterprise of two scientific unions in the area of nomenclature. And maybe their operating procedure is a good model for other organizations to follow.

INTERNATIONAL UNION OF PURE AND APPLIED CHEMISTRY

The International Union of Pure and Applied Chemistry (IUPAC), founded in 1919 after World War I as the International Union of Chemistry, immediately established commissions on nomenclature of inorganic chemistry, organic chemistry and biological chemistry. Figure 1 shows the activity of the Commission on the Nomenclature of Biological Chemistry from 1922 through 1949 [1]. G. Bertrand of France was the prime motivator for this early work, and he served as the Chaiman of this Commission for many years. The Commission met regularly at each conference of the Union. During this time the Commission worked in close contact with the Commission on Nomenclature of Organic Chemistry exchanging reports of common interest. So, IUPAC has a long history of setting standards for nomenclature in this field.

K. Loening (Editor)
The Terminology of Biotechnology:
A Multidisciplinary Problem

Year	Place of Meeting	Chairman Presiding over Discussions	Subjects Treated
1922	Lyon	E. Paterno	Preliminary work
1924	Cambridge	G. Bertrand	Bases for classification of biochemical constituents
1924	Copenhagen	G. Bertrand	Carbohydrates (glucides)
1925	Bucharest	E. Votoček	Carbohydrates (glucides)
1926	Washington	G. Bertrand	Proteins
1927	Warsaw	G. Bertrand	Glucides, Proteins, Enzymes
1928	The Hague	G. Bertrand	Fats
1930	Liége	G. Bertrand	Future organization of work
1934	Madrid	G. Bertrand	Modifications of methods of work
1936	Lucerne	Sir Arthur Harden	Enzymes
1938	Rome	G. Barger	Enzymes
1947	London	P. Karrer	Carotenoids, Amino acids
1949	Amsterdam	P. Karrer	Amino acids, Vitamins

Figure 1. Activity of the International Commission on Nomenclature of Biological Chemistry. [1]

INTERNATIONAL UNION OF BIOCHEMISTRY

On the other hand, the International Union of Biochemistry (IUB) is much more recent. It was organized in 1953 in part as a reaction to the lack of equal access to the program of the International Congress of Physiology in 1947. At about the same time, IUPAC reconstituted itself into six sections one of which would cover Biological Chemistry. Thus, many questioned the need for a separate union for biochemistry [2].

Nevertheless IUB was founded for the purpose of advancing biochemistry internationally by supporting the International Congress of Biochemistry, by coordinating research, discussion, and publications in the field of biochemistry, and by establishing cooperation among biochemical societies in adhering countries. IUB, like IUPAC, is a nongovernmental scientific organization and a member of the International Council of Scientific Unions (ICSU).

IUB was concerned with nomenclature from the beginning, particularly with enzyme nomenclature. In 1952, Professor Otto Hoffmann-Ostenhof published his famous paper "Suggestions for a More Rational Classification and Nomenclature of Enzymes". In order to bring about general acceptance of a reform of this kind he called for the "intervention of an authorative international body,..." [3].

At the occasion of the 3rd International Congress of Biochemistry in Brussels in 1955, the General Assembly of IUB decided to set up a such a commission. In 1961 the IUB Commission on Enzymes issued a report, and was dissolved. A Standing Committee on Enzymes was set up to handle comments and criticisms of the report. and a second report was issued in 1964 [4]. I don't want to go into further detail, because you will hear from Professor Edwin Webb on this topic later on.

IUPAC-IUB COMBINED COMMISSION ON BIOCHEMICAL NOMENCLATURE

At first the two unions were competitive, but this was solved in the 1960's with the creation of the IUPAC-IUB Combined Commission on Biochemical Nomenclature (CBN). In 1965 IUPAC disbanded their Commission on Nomenclature of Biological Chemistry and IUB disbanded their Standing Committee on Enzymes and the two unions jointly established a new Commission on Biochemical Nomenclature [5]. Members of the Commission were appointed, half by IUPAC, half by IUB.

A number of recommendations and tentative rules concerning abbreviations, symbols and nomenclature in biochemistry were issued by the Combined Commission. These recommendations were published by IUPAC as tentative rules in *Pure and Applied Chemistry,* and by IUB as "Recommendations 196_ " in as many as seven or eight biochemical journals. Supporting the multiple publication of the recommendations by IUB was their very strong and successful Committee of Editors of Biochemical Journals (CEBJ). The publication policy of IUPAC was much more restrictive at that time. A compilation of these recommendations was published in 1973 by IUPAC-IUB [6]. This was the first edition of the Compendium.

IUPAC-IUB JOINT COMMISSION ON BIOCHEMICAL NOMENCLATURE

In IUPAC, the new CBN was attached to the Commission on the Nomenclature of Organic Chemistry (CNOC). However, a number of factors including differing priorities, and publications policies between the two unions as well as misinterpretation by some editors as to the intent of some commission members to force compliance with the recommendations led to the necessity for some changes in the working relationship of the two unions. Thus, in 1977 CBN was disbanded. In its place, the Nomenclature Committee of IUB (NC-IUB) was set up, and the IUB-IUPAC Joint Commission on Biochemical Nomenclature (JCBN) was formed. Responsibility for enzyme nomenclature passed to NC-IUB. In IUPAC, JCBN was attached to the IUPAC Interdivisional Committee on Nomenclature and Symbols (IDCNS) where it remains today [7].

The official organization of JCBN is:

(i) There shall be a standing IUB-IUPAC Joint Commission on Biochemical Nomenclature, composed of eight regular Members, one of whom shall serve as the Chairman and one as the Secretary, and up to eight Associate Members.

(ii) The Chairman and the Secretary shall each be appointed for a term of three years by the Presidents of the two Unions, subject to whatever ratification is imposed by Union Statutes and Bylaws. JCBN may purpose names of persons suitably qualified for appointment.

(iii) Four of the regular Members shall be appointed by IUB and four by IUPAC. One of the four IUB regular Members shall be an Officer of the IUB Committee of Editors of Biochemical Journals (CEBJ). Four of the Associate Members shall be appointed by IUB and four by IUPAC.

(iv) The regular Members appointed by IUB shall also be Members of the Nomenclature Committee of IUB (NC-IUB).

(v) Candidates for membership may be proposed by the Commission and shall be appointed by the respective Unions.

(vi) The periods of service of the regular Members and of the Associate Members shall be in accordance with the Statutes and Bylaws of the appointing Union. The sum of the years of service as a regular Member and as the Chairman or the Secretary shall not exceed ten years.

Representatives from other nomenclature bodies such as the International Union of Nutritional Sciences (IUNS), and the International Federation of Clinical Chemistry (IFCC) have participated in NC-IUB and JCBN deliberations.

On the IUB side there is the Committee of Editors of Biochemical Journals (CEBJ) already previously mentioned. IUPAC has no counterpart to this Committee. CEBJ represents a unique relationship between editors and nomenclature specialists. To my knowledge there is no other field that has such a relationship. The formation of CEBJ can be traced to the controversy over NAD versus DPN. CEBJ was to supervise the work of the IUB Standing Committee on Enzymes formed in 1961.

The IUB-IUPAC (JCBN) joint enterprise has operated successfully since the mid 1970's and is renewed every 3 years. It has just been renewed in August 1989 in Lund, Sweden by IUPAC.

IUPAC-IUB PUBLICATIONS

The success of these Commissions concerning biochemical nomenclature can be attributed in some measure to its capable chairmen over the years listed in Figure 2.

Year	Name	Country
1965-1976	O. Hoffmann-Ostenhof	Austria
1977-1984	P. Karlson	Federal Republic of Germany
1985-1988	H. B. F. Dixon	United Kingdom
1989-	J. F. G. Vliegenthart	Netherlands

Figure 2. Chairmen of IUB-IUPAC (CBN) and IUB-IUPAC (JCBN)

What has come out of the joint effort? A number of topics for which recommendations have been issued are listed in Figure 3. Topics under discussion include neurotransmitters, prostaglandins and electron transfer proteins. A current bibliography of IUPAC-IUB recommendations is given in the Appendix.

Abbreviations and Symbols	Multienzyme Proteins
Amino Acids and Peptides	Natural Products
Carbohydrates	Nucleic Acid Sequences
Carotenoids	Oligosaccharide Chains
Citation of Bibliographic References	Peptide Hormones
Corrinoids	Peptidoglycans
Cyclitols	Phosphorus-Containing Compounds
Enzymes	Polymerized Amino Acids
Enzyme Kinetics	Polynucleotide Chains
Equilibrium Data	Polysaccharide Chain Conformations
Folic Acid and Related Compounds	Polysaccharides
Glycopeptides	Prenols Retinoids Steroids
Glycoproteins	Synthetic Polypeptides
Human Immunoglobulins	Tetrapyrroles
myo-Inositols	Thermodynamic Data
Iron Sulfur Proteins	Tocopherols
Isotopically Labelled Compounds	Vitamin D
Lipids	
Monosaccharides	
branched chain	
unsaturated	

Figure 3. Topics of IUPAC-IUB/JCBN and NC-IUB Recommendations

Of special mention are the NC-IUB book *Enzyme Nomenclature* [8], and the very useful compendium *Biochemical Nomenclature and Related Documents* [9]. The Compendium is a collection of documents not just from JCBN or NC-IUB, but also includes documents from CNOC and CEBJ - anything relevant to biochemical nomenclature. A document prepared by the IUPAC-IUB-IUPAB Commission on Biothermodynamics on the presentation of biochemical equilibrium data and recommendations of the International Union of Immunological Sciences for the nomenclature of human immunoglobulins are included as well. A new edition is expected any time now.

IUPAC-IUB NEWSLETTER

One last item I need to mention, the NC-IUB AND IUB-IUPAC (JCBN) Newsletter. To my knowledge, JCBN and NC-IUB were one of the first groups to pioneer this form of communicating nomenclature information. This newsletter has appeared every year since 1980. It is designed to inform the biochemical community about the work of the committees, but stops short of formal recommendations. Attention is called to features of recommendations which have been recently published or are in press. Comments on any item in the newsletter, or on any other aspect of biochemical nomenclature are invited. The "Newsletter 1989" has been published in several journals [10]. The practice of publishing newsletters has caught on with other nomenclature commissions. Both the IUPAC Commission on Nomenclature of Inorganic Chemistry (CNIC) and CNOC now issue newsletters.

I hope that this brief review of the history and recent organization of JCBN and NC-IUB will prove to be a useful starting point to consider the nomenclature problems of biotechnology.

REFERENCES

1 Courtois, J. E., "Work of Commission on Nomenclature of Biological Chemistry." *Adv. Chem. Ser.* **1953**, 83-94.

2 Slater, E. C., "The Uncertain Birthday of IUB", *Trends Biochem. Sci.* **1980**, 5(7), VII-VIII.

3 Hoffmann-Ostenhof, O. "Suggestions for a More Rational Classification and Nomenclature of Enzymes." *Adv. Enzymol.* **1952**, 14, 219-260.

4 IUB, *Enzyme Nomenclature*, Elsevier, Amsterdam, **1965**, 219 pp.

5 IUPAC, *Comptes Rendus XXIV Conference*, Prague, September **1967**, p. 106.

6 IUPAC-IUB, *Collected Tentative Rules & Recommendations of the Commission on Biochemical Nomenclature*, Reprinted by the American Society of Biological Chemists, Inc., **1973**, 117 pp.

7 "IUPAC Handbook 1988-1989", *Chemistry International* Supplement, **1988**. Blackwell Scientific Publications. p. vi.

8 IUB, *Enzyme Nomenclature,1984*, Academic Press, New York, **1984**, 650 pp.

9 IUB, *Biochemical Nomenclature and Related Documents*, The Biochemical Society, London, **1978**. 223 pp.

10 NC-IUB and IUPAC-IUB (JCBN), "Newsletter 1989" *Arch. Biochem. Biophys.* **1989**, 272(1), 262-266. *Eur. J. Biochem.* **1989**, 183, 1-4.

APPENDIX

A CURRENT BIBLIOGRAPHY OF IUPAC-IUB RECOMMENDATIONS

IUPAC-IUB, *Biochemical Nomenclature and Related Documents.* The Biochemical Society, London, 1978. 223 pp. (The Compendium).

IUB, *Enzyme Nomenclature 1984.* Academic Press, Orlando, Fla., 1984. 646 pp.
Supplement 1: Corrections and additions, Eur. J. Biochem. 1986, 157(1), 1-26.
Supplement 2: Corrections and additions, Eur. J. Biochem. 1989, 179, 489-533.
Supplement 3: Corrections and additions, Eur. J. Biochem. 1990, 187, 263-81.

"Nomenclature of tetrapyrroles." Recommendations 1986. Pure Appl. Chem. 1987, 59(6), 779-832.

"Nomenclature of steroids." Recommendations 1989. Eur. J. Biochem. 1989, 186(3), 429-458.

"Nomenclature of vitamin D." Recommendations 1981. Eur. J. Biochem. 1982, 124(2), 223-7.

"Conformational nomenclature for five and six-membered ring forms of monosaccharides and their derivatives." Recommendations 1980. Eur. J. Biochem. 1980, 111, 295-8.

"Nomenclature of branched-chain monosaccharides." Recommendations 1980. Eur. J. Biochem. 1981, 119(1), 5-8; corrections, Eur. J. Biochem. 1982, 125(1), 1.

"Nomenclature of unsaturated monosaccharides." Recommendations 1980. Eur. J. Biochem. 1981, 119(1), 1-3; corrections, Eur. J. Biochem. 1982, 125(1), 1.

"Abbreviated terminology of oligosaccharide chains." Recommendations 1980. J. Biol. Chem. 1982, 257(7), 3347-51.

"Polysaccharide nomenclature." Recommendations 1980. J. Biol. Chem. 1982, 257(7), 3352-54.

"Symbols for specifying the conformation of polysaccharide chains." Eur. J. Biochem. 1983, 131(1), 5-7.

"Nomenclature of tocopherols and related compounds." Recommendations 1981. Eur. J. Biochem. 1982, 123(3), 473-5.

"Nomenclature of retinoids." Recommendations 1981. Eur. J. Biochem. 1982, 129(1), 1-5.

"Nomenclature of iron-sulfur proteins." Recommendations 1978. Eur. J. Biochem. 1979, 93, 427-30; corrections, Eur. J. Biochem. 1979, 95, 369 and 1979, 102, 315.

"Nomenclature for multienzymes." Recommendations 1989. Eur. J. Biochem. 1989, 185(3), 485-486.

"Symbolism and terminology in enzyme kinetics." Recommendations 1981. Eur. J. Biochem. 1982, 128(1), 281-91.

"Abbreviations and symbols for the description of conformations of polynucleotide chains." Eur. J. Biochem. 1983, 131(1), 9-15.

"Nomenclature and symbolism for amino acids and peptides." Eur. J. Biochem. 1984, 128, 9-37. Recommendations 1983, "Corrections," Eur. J. Biochem. 1985, 152, 1.

"Nomenclature for incompletely specified bases in nucleic acid sequences." Recommendations 1984. Eur. J. Biochem. 1985, 150(1), 1-5.

"Recommendations for the presentation of thermodynamic and related data in biology." Pure Appl. Chem. 1986, 58(10), 1405-10.

"Nomenclature of glycoproteins, glycopeptides and peptidoglycans." Recommendations 1985. J. Biol. Chem. 1987, 262(1), 13-18.

"Nomenclature of prenols." Recommendations 1986. Pure Appl. Chem. 1987, 59(5), 683-89.

"Nomenclature and symbols for folic acid and related compounds." Recommendations 1986. Pure Appl. Chem. 1987, 59(6), 833-836.

"Numbering of atoms in *myo*-inositol." Recommendations 1988. Biochem. J. 1989, 258, 1-2.

Nomenclature and Protein Databases

Akira Tsugita

Japan International Protein Information Database (JIPID), Research Institute for Biosciences, Science University of Tokyo, 278 Noda, Japan.

INTRODUCTION

Unambiguous identification of organisms and molecules is necessary for science. Our Linnaean inheritance of latinized scientific names for biological organisms allows scientists, geographically dispersed and in different disciplines, in full confidence, to discuss the same organism. However in the rapidly evolving biological sciences and applied biotechnologies, a plethora of novel terms are accumulating. While much of this new vocabulary is indispensable in describing specific processes, novel compounds and organisms etc. many of these terms lead to confusion with existing terminology in a variety of fields. Prevailing guidelines on biological nomenclature and terminology prevent much confusion in the disciplines where they exist. And although many ambiguous terms will simply disappear through lack of usage, standardization of nomenclature should be encouraged where feasible.

Advances in the biosciences have resulted in an information explosion, particularly for biological macromolecule sequences, to such an extent that computers are the only feasible means of data storage. The almost Malthusian increase in sequence elucidation was generally unforeseen and this has caused a strain on the existing capabilities of both protein and nucleic acid sequence databases. One of the methods that has been adopted

K. Loening (Editor)
The Terminology of Biotechnology:
A Multidisciplinary Problem

to accommodate this increased workload has been for the sequence databases around the world to collaborate more closely with each other, sharing in data acquisition and dissemination etc. As the sequence databases become more integrated it will become increasingly difficult to tolerate their disparate terminologies. The efficient utilization of sequence databases requires that the nomenclature for proteins, genes, biological organisms, and scientific terminology be standardized and that these standards are followed.

There are authoritative bodies that provide guidance in many areas of nomenclature such as the International Union for Pure and Applied Chemistry (IUPAC) for bases and amino acid symbols, and the Enzyme Commission for enzyme nomenclature. Unfortunately coverage of other scientific areas is not as extensive or not applicable. This means that when novel proteins are entered into protein databases for example, the protein databases themselves in effect become the authoritative body for many areas of nomenclature and terminology. Irrespective of whether the database staff are capable of such a role, it is neither their function nor responsibility. One alternative is to have authoritative bodies for all areas of nomenclature, and have them in frequent contact with the databases.

In early 1987, the CODATA Task Group on the Coordination of Protein Data Banks (now renamed the CODATA Task Group on Biological Macromolecules) agreed to collect and exchange nomenclature and terminology between databases. The following year, CODATA organized a biological nomenclature workshop which resulted in the establishment of a commission to investigate the standardization of biological and biochemical terminology and nomenclature. This commission contains representatives from the nomenclature committees of a variety of international biological unions, protein and nucleic acid sequence databases, other biological databases, Chemical Abstracts, and experts in biomedical, microbiological, and protein nomenclature.

PROTEIN SEQUENCE DATABASES

As early as 1984, the Task Group recommended that data collection for protein sequence databases should be shared between regional centers limiting duplication of effort to a minimum. And that the datafiles of these regional centers be combined into one public domain protein sequence database. This resulted in an international collaboration of protein sequence databases called PIR-International. The databases in PIR-International are the Protein Identification Resource (PIR) at the National Biomedical Research Foundation (NBRF) in the U.S.A., the Martinsried Institute for Protein Sequences (MIPS) in the F.R.G., and the Japan International Protein Information Database (JIPID) in Japan. Each of these databases collects data in its own geographical area. JIPID for example collects data in Asia and Oceania both directly and from contributing data collection centers in Australia, India, Taiwan, Korea, and the People's Republic of China.

Due to the diverse input formats between the different sequence databases it was inconvenient to transfer sequence information between databases. The CODATA Task Group encouraged the adoption of the 'CODATA Sequence Data Exchange Format' as a means of transfer of sequence data between databases using different formats. This CODATA exchange format is quickly becoming an accepted standard in the world. An update of the Authorin program due to be released in 1990, with a Macintosh version scheduled in the same year will display protein sequences in the CODATA format.

PIR-International databases are now working towards establishing a common vocabulary of terms that will facilitate transfer of information between the databases. For example, all the PIR-International databases have agreed to follow the taxonomic classifications recommended by MIPS.

In addition, PIR-International together with the major nucleic acid databases (EMBL, GenBank, and DDBJ) are currently working on a dictionary of biological terms such as the feature table and internationally accepted standards for viruses, other biological sources gene names and chromosomal numbering.

Besides the protein and nucleic acid sequence databases, the Protein Data Bank in Brookhaven, and several novel databases such as the NMR structure database in Wisconsin, and CarbBank (the international carbohydrate database) have agreed to cite all references in the National Library of Medicine format.

The nomenclature adopted by the protein sequence databases is also being adopted by the Artificial Variant, the Biological

Activity and the Physical Property Databases, three new databases developed by PIR-International which catalogue artificially created novel proteins and the biological activities and physical properties of proteins respectively. These databases are being included into the PIR-International network to provide the user with a more comprehensive biological information database.

PROTEIN NOMENCLATURE

Although there is no clear distinction between the terms protein or polypeptide, PIR-International take the traditional view that sequences of amino acids with a molecular mass of less the 10,000 be referred to as polypeptides. Amino acid sequences are generally called 'chains'. The appellation 'subunit' is avoided unless the molecule is divided into subunits which themselves consist of several chains.

A protein chain may contain several sequences that possess distinct biological properties in vivo. We refer to each of these sequences as functional domains. Such a protein may undergo post-translational cleavage to yield several proteins each containing one or more functional domains. These proteins are called a polyprotein. Proteins that remain functionally inert, i.e. in an immature state, unless post-transcriptionally modified or cleaved is termed a pro-protein. Also a precursor to a pro-protein may exist, which is called a pre-pro-protein, or precursor protein.

When an amino acid sequence is determined from the nucleic acid sequence without direct evidence of the existence of the molecule, the sequence is listed in the database as a 'Hypothetical protein'. The terms URF and ORF are not used. Also although protein molecular mass is usually written as kDa, in the database it is written only as K. The genomic region of the hypothetical protein is written in parentheses. Therefore an E. coli entry may be written as:

Hypothetical 12K protein (thrC 3'region)

Enzymes are written as recommended by the Enzyme Nomenclature Committee in the Enzyme Nomenclature (1984) and the following supplements. The enzyme name is immediately followed by the EC number in parentheses. The EC name may refer to a general class, a specific type, or even to a chain within a molecule or domain. Therefore the EC number may appear in various places in the protein title.

Other conventions include the use of adjectival forms for the protein name. Examples of these forms are:

cytosolic NOT cytoplasmic

hepatic NOT liver

cardiac NOT heart

also: pancreatic, salivary, intestinal, embryonic, gonodal, cystskeletal, etc., but when there is no convenient adjectival form: brain sperm, muscle , etc., may be used. The terms 'type' and 'isozyme' are not used unless they are an integral part of the principal name.

The title line of an entry contains the protein name, a dash and the name of the organism. Immediately following the organism name is the type of organelle e.g. Paramecium aurelia mitochondrion.

However in the Species data item contains the name of the organelle followed by the organism name e.g., mitochondrion Paramecium aurelia.

Commonly accepted abbreviations are used in the database. For example, immunoglobulins are always abbreviated to 'Ig' followed by the type of chain and the region of the chain.

Ig mu chain C region - Golden hamster.

If there are any international agreements on nomenclature such as that for P450 gene nomenclature (DNA), the PIR-International databases will endeavor to follow the recommended guidelines.

Without standardization of nomenclature and terminology, communication between the biopolymer databases may become increasingly inhibited. Fortunately this problem has been realized. Several biological macromolecule databases and authoritative bodies for the nomenclature of several biological and chemical disciplines are cooperating to alleviate the problem.

The Nomenclature of Altered Proteins

John A. Secrist III

Kettering-Meyer Laboratory, Southern Research Institute
P. O. Box 55305, Birmingham, AL 35255-5305

Abstract: The scientific literature contains an increasing number of reports of proteins altered in many different ways. It is becoming evident that a nomenclature system needs to be implemented to bring some order to this research area. Comments and suggestions relating to what such a system should include, focusing on specific recommendations in certain areas, are presented in this report.

INTRODUCTION

Biotechnology nomenclature has many facets to it, as can be seen by this volume. Some of the problems that require solutions are indeed formidable, and the volume of information that needs to be organized is quite large. An aspect of biotechnology nomenclature that appeared to be both in need of attention and somewhat manageable was that of altered proteins. The widespread use of the techniques of molecular biology has resulted in the routine reporting of many proteins that are altered in a variety of ways, and any nomenclature system would need to be flexible enough to encompass all of these alterations as well as others that might occur in the future. A further concern is the sheer volume of information that needs to be cataloged, by Chemical Abstracts or other organizations, in order to properly and completely represent these altered proteins. If the primary structure of each altered protein, even with only a point mutation, needed to be recorded, the memory space necessary is enormous. In order to initially determine whether anything beyond what is currently available was deemed desirable by scientists in the area, the Committee on Nomenclature of the American Chemical Society formed a subcommittee to look into the question [1]. Letters were sent to many researchers around the world who are actively engaged in research in some way connected with altered proteins. The responses received were unanimous in saying that some system was needed, and many comments were made, including the mention of 1) situations that should be covered by a nomenclature system, 2) specific recommendations for how to deal with certain types of altered proteins, 3) areas of concern, and 4) broader considerations. On the basis of these responses, the types of variant proteins that must be accommodated by a nomenclature system include: 1) point mutations, 2) double and multiple mutations, 3) site-directed sequence changes, 4) insertions, 5) deletions, 6) hybrid proteins. The purpose of this report is to present a status report on suggestions for a nomenclature system for altered proteins and to identify problems and questions that need further consideration. A excellent commentary appeared in early 1988 on this same topic, and the reader is referred to it for added perspective [2].

K. Loening (Editor)
The Terminology of Biotechnology:
A Multidisciplinary Problem

As a beginning, consideration needs to be given to the most recent existing nomenclature recommendations as to how well or poorly they fit at least some of the needs described above, and in terms of what situations are not covered by them. The applicable guidelines are the 1983 recommendations of the IUPAC-IUB Joint Commission on Biochemical Nomenclature (JCBN), entitled "Nomenclature and Symbolism for Amino Acids and Peptides" [3]. Modifications covered include insertions, deletions, replacements, as well as side-chain and other types of modifications. Square brackets are used to indicate replacement, and the terms endo and des are used for insertion and deletion, respectively. Replacement sites are designated by superscripts with an amino acid number in the peptide chain. Three letter codes are used for the amino acids that make up the peptide chain, and the unmodified peptide is used as the parent name, and is included in the name of each modified peptide. Multiple changes would be extremely inconvenient with this system, and large insertions or deletions would also be unwieldy. As can be seen, the system is not very streamlined, and makes the description of multiple changes a cumbersome task. Thus, it seems that a flexible and concise system for proteins is needed, one that will be adopted for general use because it is a clear and convenient way to present the information.

In the development of a system, various other factors need to be considered. For example, how is the designation of the parent protein upon which any modifications are made to be handled? Should all of the characters used by present on a standard computer keyboard? When available, should any additional data on the protein secondary structure be included? What type of numbering system for the protein should be adopted?

POINT MUTATIONS AND MULTIPLE MUTATIONS

The simplest situation to deal with is a point mutation in a protein. Systems that were suggested are listed below for the replacement of an arginine (Arg, R) to a phenylalanine (Phe, F) at position 70 of a protein. It was generally agreed that one-letter codes, now being used

Arg(70)Phe
R(70)F
R70F
RF70
F70

routinely by scientists, should be used as the most concise descriptors for amino acids, thus allowing a focus on the last four suggestions. The parentheses probably are unnecessary, also, thus saving two characters. The advantage of the RF70 system is that a series of mutations can be described in linear fashion somewhat more clearly without the need for the insertion of a semicolon or comma between mutation descriptors. A disadvantage to this system is that it could result in confusion with other types of nomenclature. The last system provides less information, but still would allow full knowledge of the replacement because the parent protein (and also and appropriate literature reference) would also be mentioned, and therefore the amino acid at position

70 could be determined from that structure. Most responders felt that the addition of only one character to indicate what amino acid has been replaced was warranted to allow a reader immediate access to this important information without having to refer back to the parent protein structure. Inclusion of the residue that is replaced also provides an internal check on the numbering system.

When protein secondary structure information is available, a system such as the following was suggested. In this system E11 describes the eleventh residue in helix E, and 63 is the

Val E11β-Ile

number in helix E, and 63 is the number in the amino acid sequence. Some consideration would need to be given to a convenient way to use one-letter codes, and of course most proteins would not be characterized to the extent necessary to add the extra information.

Double and multiple mutations, building upon the above suggestions for single mutations, might then be done as shown below. Most scientists recommended the inclusion of some

R70F/A126K	A = alanine
R70F;A126K	K = lysine
R70F,A126K	D = aspartic acid
RF70AK126	S = serine

R70F, A126K, F181A, D202S...

separation mark, those shown being a diagonal line, a semicolon, and a comma. If one uses the system with both single-letter codes in front of the amino acid number, then a system without such separations can be used, but would be somewhat more difficult to follow for a string of changes. This system is certainly is the most space efficient of those suggested. Any of the above systems could readily be used for multiple mutations, and one example, using commas as separators, is shown above.

INSERTIONS AND DELETIONS

Single residue deletions, resulting in a protein of one less amino acid, might be designated by one of the systems shown below, exemplified by the deletion of the arginine residue at position 70. These systems are all as short as or shorter than the existing peptide recommendations,

ΔR70

δR70

dR70

desR70

delR70

which use the term des. The top three, with single character codes for deletion, would seem to be the best possibilities. If one of the single character codes is chosen, then multiple deletions could be designated by a system similar to that shown above for replacements (dR70;dA126;...). The deletion of the entire region from arginine 70 to alanine 126 might be shown in the following manner.

dR70-A126

Suggestions for insertions are a lengthier because the residue after which the insertion comes must be designated. Some possibilities are shown below for the insertion of a glutamic acid after arginine 70. In a fashion similar to that shown for the deletion of an entire

R70iQ
R70insQ
R70ιQ

region, the insertion of a series of amino acids might be done as shown below for the insertion of an eight amino acid sequence after arginine 70.

R70iAKLLNQIA

SITE-DIRECTED SEQUENCE CHANGES

Site-directed sequence changes can involve a relatively few amino acids or a large number, and simple replacement nomenclature, as shown above, might be used in situations involving only a small number of amino acids. For longer sequences some better system might be considered. When a particular sequence of a parent protein in removed and another sequence is substituted, the substituted sequence need not be the same length as that deleted, but might be longer or shorter. Several approaches to nomenclature for these situations are given below. In the first example, the residues that have been removed from the parent protein are not mentioned, but

98:110→ASNSSNADTKV
(YWTSQPNV)FFSAMPLLG70

rather only those that are inserted in the altered protein. It should be noted that in the example fewer amino acids are inserted than were removed. In the second example, an eight amino acid sequence beginning at residue 70 of the parent protein is removed, and a nine amino acid sequence is inserted in its place. This nomenclature has the advantage of allowing ready examination of the sequence that has been replaced, but has the disadvantage of taking up more space in the process. The question that needs consideration is whether for such a large change it is worthwhile to include the deleted sequence, or whether that information is not important enough to be included as primary information, but should rather be left available to scientists through the reference to the sequence of the parent protein. Certainly such information would seem more important for one or several point mutations than for a large change such as shown in the above examples.

HYBRID PROTEINS

Hybrid proteins are made up of pieces of two (or more) parent proteins joined together. One possibility for describing such altered proteins is to write out the new sequence as a series of insertions and deletions of partial sequences from the parent proteins. One way to accomplish this goal is shown below. In this system, the B parent protein sequences are distinguished from

ABAB = Q1R10 N15T20 C15K22 N27N33

the A parent protein sequences by the underlining. Thus the hybrid protein is made up of the ten amino acid sequence Q1 through R10 from the A parent protein followed directly by the five amino acid sequence N15 through T20 from the B parent protein, and so on. Another possibility is to use italics for one of the parent proteins. A somewhat different system is presented below. In these examples, the residues from one protein are given in italics (*protein b*), while the

[*i(50-70)*71]*protein b*/protein a

[*(72-89)*(75-90)]*protein b*/protein a

residues for the other protein are given in normal print. This system reduces the chance for confusion relative to which sequences come from which protein. In the first example, residues *50-70* of *protein b* are inserted after residue 71 of *protein a*. In the second example residues *75-90* of *protein a* are replaced by residues *72-89* of *protein b*. The residues to be replaced might also be put as superscripts, though this system would be less satisfactory from an ease of recording standpoint. Multiple changes could easily be denoted in series in the same manner.

NUMBERING OF PROTEIN CHAINS

Accurate numbering of protein chains is obviously critical to the proper designation of what residues are replaced and where insertions occur. Several comments or suggestions were made on this topic, and it appears that the most satisfactory answer may depend upon the type of research that a scientist is pursuing. One responder suggested that the parent sequence in one-letter codes should always be given in each publication, thereby making any numbering scheme obvious. In some scientific areas, it is obviously highly desirable to keep certain numbers associated with certain amino acids in a given protein. For example, there are 28 lactate dehydrogenase sequences, all different lengths, and it would seem desirable for the active site histidine always to be H195. Attempting to incorporate this outcome into any broadly based numbering system would be extremely difficult, but if a parent was numbered within a scientific paper, then at least internal comparisons would present no problem. An interesting suggestion that would aid in keeping key residues with the same number has been previously pointed out [2], and that is to use the designation -1 for the initiator methionine. Another method that might be used for internal additions is to use small letters after the number (58a, 58b, etc.) for amino acids after which there have been insertions. The problem of deletions relative to keeping the same numbers for key residues would also require some sort of innovative solution of general applicability.

Another suggestion, another solution to the problem addressed by the -1 suggestion above, was to always begin numbering of the protein chain with the AUG of the messenger RNA as position 1, rather than the N-terminal amino acid. This end can change depending upon the level of expression, which tissue the protein was isolated from, or whether it is a processed or non-processed form of the protein. This system would be broadly applicable, but might not be favored by scientists who may be interested in active site comparisons such as that mentioned in the above paragraph. Perhaps a key question is whether a nomenclature system for altered proteins needs to include a numbering system. It may prove best to establish conventions for dealing with various alterations, and let scientists set their own numbering systems depending upon what the needs and limitations are in a particular research area.

PARENT PROTEIN

Because of the techniques now available, parent proteins may be natural or may already be altered in some way. How then should the parent protein upon which changes are being made be identified? A number of suggestions were made. One that was mentioned above is to write out the parent sequence for each altered protein in one-letter codes in each publication. This approach would be quite clear, though taking up significant journal space. A corollary suggestion is to define the parent by a specific literature reference to the protein. Certainly this step would be necessary whether or not the parent sequence was written out. When available, the abbreviation from the Brookhaven Protein Data Bank could be used. Two problems are immediately evident with this approach. First, only a relatively few proteins have had three-dimensional structures determined, and second, there are mistakes in the data bank that are gradually corrected, and a date would need to be specified with each reference. Establishing some sort of name for parent proteins could also be a challenge, depending upon how the parent was derived. One suggestion was to use four capital letters for proteins specified by known genes (e.g., GCN4, TRPE), and if the gene is not known, use GP45, etc. Degrees of complexity begin to enter when the "parent" for a particular piece of research is an already modified protein. As previously suggested [2], it may be that common names will be needed once a clearly identifiable name has been given for a particular protein.

GENERAL COMMENTS

A number of comments were made by responders with the aim of facilitating the development of a meaningful nomenclature system. Certainly the system should not be too unwieldy or it will not be readily adapted by scientists. It should be logical and reasonably easy to remember. A system that is so cumbersome that no one will adhere to it will represent just a waste of the developers time. The system must be flexible enough to handle all situations that may develop in the future. One significant complication in that regard is the question of how to handle post-translational modification of proteins. This question is touched on in the earlier commentary [2], and it is clear that this area represents an almost separate issue in terms of nomenclature development that must, however, be integrated into any altered protein nomenclature. A reasonable question to ask is whether this topic should be included in one effort to develop a usable nomenclature system, or should it rather be treated separately after a system is developed to deal with the amino acid sequences? An important point for ease of future use is that any symbols should be available on a standard computer keyboard, though the availability of almost any symbol through readily available software packages appears to make this point less of a problem. Any new system should conform to existing IUPAC-IUB nomenclature recommendations as much as possible. Obviously, if divergence is required, new solutions should be implemented; however, building upon the current recommendations is an obvious way to begin the process.

FUTURE DIRECTIONS

All responders felt that a system was needed to bring some order to the nomenclature of altered proteins. It is the feeling of this subcommittee that a panel of scientists that will use the system should be convened to begin to develop the details of an appropriate system. Optimally, the process might be conducted under the auspices of IUPAC and IUB, though perhaps the ACS and other relevant bodies could also be parties to it. Clearly, based upon the key need to manage and record the information developed and published in research along these lines, Chemical Abstracts Service needs to be involved in the process. The development of a nomenclature system is a difficult, time-consuming, and thankless task, but all of the participants surveyed felt that the time had come to proceed as rapidly as possible.

ACKNOWLEDGMENT

Thanks are due to the following scientists for their willingness to respond to our request for input on the question of a nomenclature system for altered proteins: J. J. Holbrook, University of Bristol; J. Kirsch, University of California, Berkeley; M. F. Perutz, MRC; Michael Smith, University of British Columbia; G. Winter, MRC; C. Yanofsky, Stanford University; S. Altman, Yale University; C. G. Brouillette, University of Alabama at Birmingham; C. S. Craik, University of California, San Francisco; H. R. Kaback, Roche Institute of Molecular Biology; T. A. Kunkel, NIEHS; R. A. Owens, USDA; H. K. Schachman, University of California, Berkeley; G. Vasanthakumar, Southern Research Institute; D. M. Watterson, Vanderbilt University. The support of the ACS Committee on Nomenclature was also much appreciated.

REFERENCES

1 The subcommittee was comprised of Drs. John A. Secrist III and George Gorin (Oklahoma State University) from the ACS Committee on Nomenclature, and Dr. Bruce Benjamin of Chemical Abstracts Service
2 Wetzel R (1988) Protein Engineering, 2:1
3 "Nomenclature and Symbolism for Amino Acids and Peptides, Recommendations 1983" (1984) Eur J Biochem 138:9; (1985) J Biol Chem, 260:14

Nomenclature and Representation of Biopolymer Sequences by Chemical Abstracts Service

Joy E. Merritt, Bruce M. Benjamin, and William F. Hackett

Chemical Abstracts Service,[®]
P.O. Box 3012, Columbus, Ohio 43210

Abstract: Currently there are over 120,000 protein sequences and over 35,000 nucleic acid sequences in the Chemical Abstracts Service (CAS) machine and manual files. Approximately 20,000 new sequences are reported each year. Both the current number of existing sequences and the rapid increase in the number of new sequences being reported require maintaining and developing a nomenclature system to keep up with advances in the field. Clear, unambiguous and unique names are needed in order to keep track of and to report biopolymer sequence information in abstracts, databases, and in indexes. The CAS naming and representation of amino acids, peptides, and nucleic acids including those with both short and long sequences are presented. Problems encountered with altered sequences of natural products, and those with partial sequences or with undefined or ambiguous groups are discussed.

BIOPOLYMER SEQUENCES

Peptides and proteins which are made up of amino acid sequences, and nucleic acids which are made up of nucleotide sequences, are biopolymer sequences. Peptides and proteins are biological polymers with an amide repeating unit forming their backbone (Figure 1). They are composed of amino acids linked end to end to form a long chain. They have a highly uniform backbone with various amino acid side chains attached at a fixed interval along the chain. They serve a broad range of biological functions including catalysis (enzymes), transport (hemoglobin), storage and nutrition (plant seed proteins), structure (collagen in skin), immune response (antigens and antibodies), and regulation (hormones such as insulin). Frequently, a protein is identified by its activity before its amino acid sequence is known, and *CA*[®] captures such information for its indexes. However, the discussion that follows is limited to biopolymers of known sequences.

Nucleic acids are biopolymers consisting of a backbone of alternating ribose or deoxyribose sugar and phosphate groups (Figure 2). Attached to each sugar is a purine or a pyrimidine base. The principal features that distinguish one nucleic acid from another are the order in which the specific purine and pyrimidine bases involved occur along the chain, the length of the chain, and whether ribose (RNA) or deoxyribose (DNA) is used to form the backbone.

K. Loening (Editor)
The Terminology of Biotechnology:
A Multidisciplinary Problem

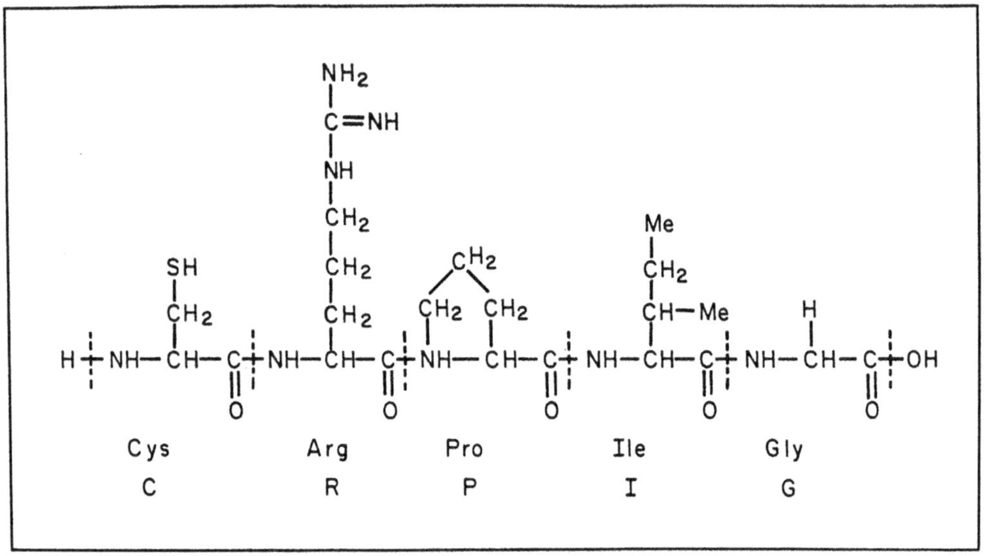

Figure 1. Peptides/Proteins

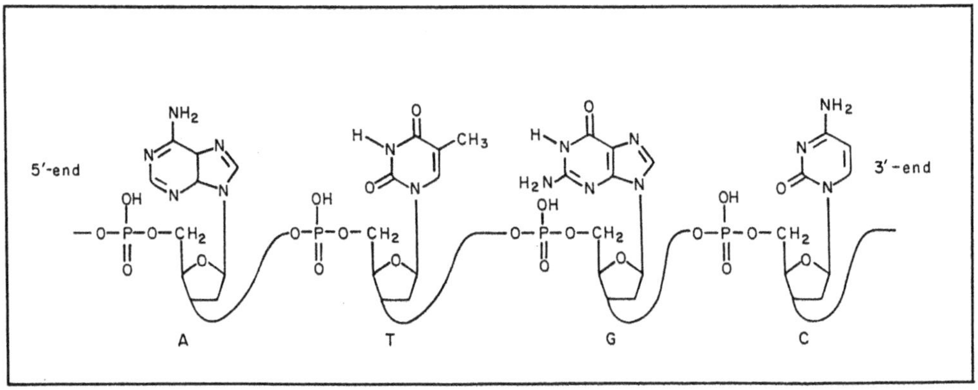

Figure 2. Nucleic Acids

SCOPE OF THE CAS NOMENCLATURE TASK FOR BIOPOLYMER SEQUENCES

Chemical Abstracts Service (CAS) is an electronic database producer and publisher of abstracts and indexes in the fields of chemistry and chemical technology. In accord with its mission to abstract and index the world's chemical literature, CAS has created a database of chemical information linked to bibliographic references. From this database, abstracts and indexes are published, and CAS ONLINE® is produced. The CA abstracts and indexes date back to 1907. The building of the CAS electronic database began in 1967, and it now contains over 9 million chemical substance records linked to 8 million bibliographic citations. This also includes records prior to 1967 that have been added retrospectively.

CAS Literature Coverage

CAS monitors approximately 10,000 titles a year from scientific and technical journals, conference and symposium proceedings, government reports, and science books. Roughly a half million documents including 85,000 patents are selected from those titles for abstracting and indexing. These articles come in 50 languages from 150 different countries. The CAS database includes protein registrations from over 1100 journals and nucleic acid registrations from over 800 journals. By comparison, the largest DNA database, GenBank, covers roughly 100 to 200 journals.

Number of Biomacromolecular Sequences in CAS Database

CAS covers virtually all types of protein and nucleic acid sequences including natural and synthetic peptides/proteins; natural and synthetic nucleic acids; translated proteins; sequences from patents, chemically modified and analog sequences, and genetically engineered sequences. Other sequence databases (GenBank/EMBL, NBRF/PIR, SwissProt) are genetic databases in the sense of their selection, coverage and organization. The CAS database is a chemical database tied to bibliographic references. This results in some unique aspects of coverage and terminology. For instance, the CAS database for proteins is not limited to the standard twenty α-L-amino acids, but documents uncommon amino acids, β, γ and retro linkages, and it allows for expression of organic chemical substitutions. For nucleic acids, variations in the purine and pyrimidine bases have been observed and are documented.

Thus, in comparison, the CAS database is very large. Currently there are over 120,000 protein sequences and over 35,000 nucleic acid sequences in the CAS machine and manual files. There are over 150,000 biopolymer sequences named and indexed by *CA*. Approximately 20,000 new biopolymer sequences are reported each year.

CURRENT NOMENCLATURE RECOMMENDATIONS AND PRACTICES

IUPAC and IUB Recommendations

With regard to naming peptides the IUPAC/IUB Joint Commission on Biochemical Nomenclature (JCBN) in 1983 issued recommendations for the "Nomenclature and Symbolism for Amino Acids and Peptides" [1]. The CAS nomenclature of natural peptide derivatives is based on these recommendations. Symbols for both a one-letter code and a three-letter code for designating the 20 common amino acids are codified (Figure 3). Recommendations for synthetic polypeptides have also been issued [2].

With regard to naming nucleic acid sequences, two major IUPAC and IUB documents have been published [3,4].

One-letter symbol	Three-letter symbol	Amino acid
A	Ala	alanine
B	Asx	aspartic acid or asparagine
C	Cys	cysteine
D	Asp	aspartic acid
E	Glu	glutamic acid
F	Phe	phenylalanine
G	Gly	glycine
H	His	histidine
I	Ile	isoleucine
K	Lys	lysine
L	Leu	leucine
M	Met	methionine
N	Asn	asparagine
P	Pro	proline
Q	Gln	glutamine
R	Arg	arginine
S	Ser	serine
T	Thr	threonine
V	Val	valine
W	Trp	tryptophan
X	Xaa	unknown or 'other' amino acid
Y	Tyr	tyrosine
Z	Glx	glutamic acid or glutamine (or substances such as 4-carboxyglutamic acid and 5-oxoproline that yield glutamic acid on acid hydrolysis of peptides)

Figure 3. IUPAC-IUB Symbols for Amino Acids [1].

CAS NOMENCLATURE PRACTICES FOR PEPTIDES AND PROTEINS

Five types of names comprise the core of CAS peptide nomenclature practices: (1) Stereoparent and Stereoparent Analog Names, (2) Peptide Systematic Names, (3) Amino Acid Sequence Names, (4) Cyclic Peptide Names, and (5) Protein Names. Each of these types will be illustrated below with actual examples.

Peptide Stereoparent and Stereoparent Analog Names

The CAS Registry file contains over 120,000 sequences of proteins and peptides. The nomenclature of these compounds reflects considerations of biological source and activity, and chain length.

Naturally occurring biologically active linear and cyclic peptides (hormones, antibiotics, etc.) containing six through fifty amino acid residues, for which the complete sequence has been determined, are assigned trivial "peptide stereoparent names" (Figure 4) which reflect their biological activity and source. Since these peptide stereoparents are biologically important, every attempt is made to assign to other peptides possessing similar structures (whether of natural or synthetic origin) "analog names" (Figure 5) based on the peptide stereoparents which describe the specific structural relationship, when that relationship can be recognized.

CA INDEX NAME: **Adipokinetic hormone I (Locusta migratoria)**

SEQUENCE: (Limit of 6 to 50 amino acid residues)

H-5-oxo-L-Pro-L-Leu-L-Asp(NH_2)-L-Phe-L-Thr-L-Pro-L-Asp(NH_2)-L-Trp-Gly-L-Thr-NH_2
1 2 3 4 5 6 7 8 9 10

Figure 4. Peptide Stereoparent

CA INDEX NAME: **Adipokinetic hormone I (Locusta migratoria), 2-L-valine-3-L-alanine-5-L-serine-7-glycine-**

SEQUENCE: (Limit of 4 to 50 amino acid residues)

H-5-oxo-L-Pro-L-**Val**-L-**Ala**-L-Phe-L-**Ser**-L-Pro-**Gly**-L-Trp-Gly-L-Thr-NH_2
1 2 3 4 5 6 7 8 9 10

Figure 5. Stereoparent Analog Name

Recognition is the key word for the practical application of this policy. In many cases, the structural similarity is not immediately apparent -- particularly since the file of peptide stereoparent structures now present in the *Chemical Abstracts Index Guide* contains more than 500 entries. As an aid in this recognition, computer files of these stereoparents are maintained, together with programs for searching these files.

If a peptide can be related structurally to a known peptide stereoparent by any combination of changes (replacement, deletion, or insertion of amino acids) which totals less than half of the number of amino acids in the stereoparent, that peptide is assigned an analog name on the basis of the corresponding stereoparent. Since deletions are among the allowed changes, a successful candidate for such a "peptide stereoparent analog name" must generally contain at least four amino acids.

Peptide Systematic Names and Amino Acid Sequence Names

If a peptide does not qualify for assignment of a peptide stereoparent analog name as described above, the index name assigned depends, to a large degree, on the number and type of amino acids in the chain or ring.

In general, linear peptides containing two through twelve amino acid residues are assigned "peptide systematic names" (Figure 6), if the C-terminal residue is a carboxylic acid. However, if the C-terminal residue is an amide the linear peptide

CA INDEX NAME: **L-Alanine,**
N-[N-[N-[N^2-[N-[N-(N^2-L-lysyl-L-lysyl)-L-seryl]-L-seryl]-L-lysyl]-L-alanyl]-L-alanyl]-L-alanine

SEQUENCE: (Limit of 2 to 12 amino acid residues)

Figure 6. Peptide Systematic Name

CA INDEX NAME: **L-Glutamine,**
L-alanyl-L-threonyl-L-tyrosyl-L-leucyl-L-asparaginyl-L-arginyl-L-α-glutamyl-L-prolyl-L-glutaminyl-L-leucyl-L-lysyl-L-alanyl-L-prolyl-L-threonyl-L-prolyl-L-histidyl-

SEQUENCE: (More than 12 amino acid residues)

CA INDEX NAME: **L-Phenylalaninamide,**
L-α-aspartyl-L-prolyl-L-phenylalanyl-L-leucyl-L-arginyl-

SEQUENCE: (2 to 12 amino acid residues and a C-terminal amide)

CA INDEX NAME: **Glycine,**
N-acetyl-L-alanyl-L-2,4-diaminobutanoyl-L-isoleucyl-

SEQUENCE: (2 to 12 amino acid residues and Abu or similar residue)

Figure 7. Amino Acid Sequence Names

is assigned an "amino acid sequence name" (Figure 7). Nevertheless, if certain less common α or β amino acids named by organic systematic nomenclature (such as 2-aminobutanoic acid) occur in a nonterminal position within a linear peptide, or if a disulfide bond joins two peptide chains, an amino acid sequence name is required, regardless of C-terminal functionality. A "peptide systematic name" differs from an "amino acid sequence name" in that in the former the specific linkage of one amino acid to another is specifically stated, whereas in the latter, it is assumed to be linked through the α nitrogen.

Cyclic Peptides and Cyclic Peptide Names

Cyclic peptides containing fewer than four amino acids are named systematically by the rules of general organic nomenclature (Figure 8). Cyclic peptides containing rings of four or more amino acids are assigned "cyclic peptide names" (Figure 9), which resemble "amino acid sequence names".

CA INDEX NAME: **2,5-Piperazinedione, 3-[(4-hydroxyphenyl)methyl]-6-(1H-imidazol-4-ylmethyl)-, (3S-cis)-**

SYNONYM: Cyclo(L-tyrosyl-L-histidine)

STRUCTURE: (Ring of fewer than 4 amino acid residues)

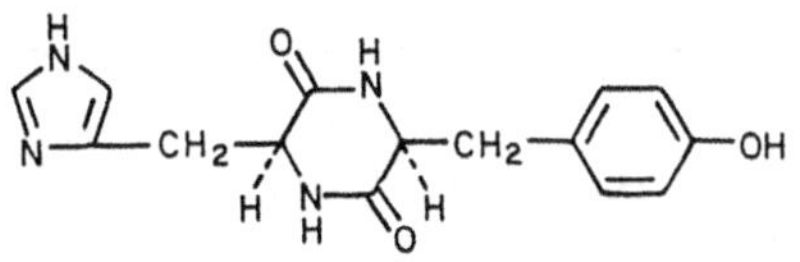

Figure 8. Cyclic Peptide Named Systematically

CA INDEX NAME: **Cyclo(L-arginyl-L-tyrosyl-L-isoleucylglycyl-L-seryl)**

STUCTURE: (Ring having 4 or more amino acid residues)

Tyr-Ile-Gly-Ser-Arg

Figure 9. Cyclic Peptide Name

Protein Names

Proteins are arbitrarily defined as naturally occurring peptides (or derivatives) that contain more than fifty amino acids. Nomenclature policies for proteins are similar to those for other peptides with trivial names, but the naming of proteins is complicated by the frequent occurrence of subunits (*i.e.*, separate chains, covalently or noncovalently bonded to other chains), proteins that are precursors in the cell of other proteins, and proteins that have prosthetic groups (carbohydrates, lipids, metals, etc.). In general, the specific protein structure registered is usually the "protein moiety" (*i.e.* the amino acid sequence without inclusion of any prosthetic groups). However, proteins along with various types of prosthetic groups can and have been registered.

As with the peptide stereoparent analogs described earlier, protein structures are preferably assigned index names which reflect their biological activity and source -- either in their own right or as analogs of other proteins (Figure 10). Here the requirement that an analog fragment must contain more than half of the

CA INDEX NAME: **Acetyltransferase, formate (Escherichia coli subunit protein moiety reduced)**

CA INDEX NAME: **Tumor necrosis factor (human reduced)**

CA INDEX NAME: **8-157-Tumor necrosis factor (human reduced), 32-L-glutamic acid-**

SEQUENCES: (More than 50 amino acid residues)

Figure 10. Protein Names

total amino acids in the reference protein is relaxed; if a protein fragment contains more than fifty amino acids, it is named as such (by prefixing a residue-number range to the name of the total reference protein), regardless of the length of the reference protein. As before, however, the total number of changes from the parental compound must be less than half of the number of amino acids in that parental compound, whether that compound is a complete protein or a protein fragment.

If a protein structure does not qualify for assignment of a "protein name" or a "protein analog name" (or if its relationship to a naturally occurring protein cannot be determined), it is assigned an "amino acid sequence name". Naturally occurring sequences can be functionally described by a "homograph" data element which indicates the biological source and coding or functional properties. Controlled vocabulary for the biological source is based upon taxonomic standards; the description of coding properties is based upon the nomenclature of the protein/peptide product or the gene designation.

In index names, the CAS-preferred names are used for registered sequences. When a registered sequence is based on a general subject, CAS-preferred taxonomic names are used. Additional specificity can be added to the index names by inclusion of designations of strain and cell line. The ultimate degree of specificity possible with descriptive English language terms (as opposed to systematic names based on structure) can be provided by inclusion of clone designations, when possible (author-supplied) and desirable. In names of proteins, "protein moiety" indicates the presence of other chemical groups (e.g. oligosaccharides, prosthetic groups) that are present in the molecule, but not depicted in the structure. The word "reduced" indicates the presence of two or more cysteine residues with no or unknown disulfide bridges.

CAS NOMENCLATURE PRACTICES FOR NUCLEIC ACIDS

As stated earlier, the CAS Registry File contains over 35,000 sequences of nucleic acids and oligonucleotides. IUPAC standards for the naming of nucleotides are in universal usage for the basic five nucleotides [3,4]. Since heterogeneity or ambiguity is common with nucleic acid sequences, widely accepted standards also exist that represent any combination of these nucleotides (Figure 11). Many of the modified nucleotides are also represented by commonly accepted symbols.

Symbol	Meaning	Origin of designation
G	G	Guanine
A	A	Adenine
T	T	Thymine
C	C	Cytosine
R	G or A	puRine
Y	T or C	pYrimidine
M	A or C	aMino
K	G or T	Keto
S	G or C	Strong interaction (3 H bonds)
W	A or T	Weak interaction (2 H bonds)
H	A or C or T	not-G, H follows G in the alphabet
B	G or T or C	not-A, B follows A
V	G or C or A	not-T (not-U), V follows U
D	G or A or T	not-C, D follows C
N	G or A or T or C	aNy

Figure 11. IUB Symbols for Nucleic Acids [3,4].

The current limitations to machine registration in the CAS Registry database apply to nucleic acids larger than about 13 nucleotide residues. The vast majority of naturally occurring nucleic acids and functionally interesting nucleic acid fragments are much larger. Thus, a consistent, recognizable nomenclature system is essential. Three types of names comprise the core of the CAS nomenclature practice for nucleic acids: (1) Nucleic Acid Systematic Names, (2) Single Letter Codes, (3) Names Containing Homographs. The type of name selected is dependent on the size and the biological function of the nucleic acid.

Nucleic Acid Systematic Names

The smallest (20 or fewer nucleotide units) of the nucleic acids can be named using radicals derived from the individual nucleotide names (Figure 12). The positions of the ester linkages are specified by the appropriate locants separated by an arrow. This scheme is very analogous to that for peptide systematic names.

CA INDEX NAME:

Guanosine,
2′-deoxyguanylyl-(3′→ 5′)-thymidylyl-(3′→ 5′)-2′-deoxyadenylyl-(3′→ 5′)-2′-deoxyguanylyl-(3′→ 5′)-2′-deoxyadenylyl-(3′→ 5′)-2′-deoxyadenylyl-(3′→ 5′)-2′-deoxycytidylyl-(3′→ 5′)-2′-deoxyadenylyl-(3′→ 5′)-thymidylyl-(3′→ 5′)-thymidylyl-(3′→ 5′)-thymidylyl-(3′→ 5′)-thymidylyl-(3′→ 5′)-2′-deoxyadenylyl-(3′→ 5′)-thymidylyl-(3′→ 5′)-2′-deoxycytidylyl-(3′→ 5′)-2′-deoxy-

SEQUENCE: (20 or fewer nucleotide units)

Figure 12. Nucleic Acid Systematic Name

Deoxyribonucleic Acid and Ribonucleic Acid Names

Nucleotide sequences containing more than 20 units are named as a "Deoxyribonucleic acid" or a "Ribonucleic acid". In many synthetic sequences or fragments of biological nucleic acids, no defined function is known. These are named by a format which indicates the structure using the one-letter code for each nucleotide unit (Figure 13). This nomenclature scheme is unique to nucleic acids within the CAS Registry File.

Nucleic Acid Homograph Names

The third general scheme for nucleic acid nomenclature is again very similar to that used for larger proteins. Thus, the so-called "homograph" data element contains information describing the biological source of the nucleic acid and its

CA INDEX NAME:

Deoxyribonucleic acid,
d(G-C-C-A-G-A-G-C-T-G-T-T-G-A-G-G-G-A-A-G-G-A-A-A-C-C-T-C-
G-A-G-G-A-G-G-C-A-G-C-A-A-T-G-G-T-G-G-A-T-G-T-T-G-G-A-T-
C-G-A-G-G-T-G-G-A-G-G-C-T-A-A-C-C-A-G-T-A-C-A)

SEQUENCE: (More than 20 nucleotide units)

Figure 13. Nucleotide Sequence Name

coding or functional properties (Figure 14). For informational nucleic acids, such as genomic DNA or viral RNA, or messenger RNA, the coding properties are described by the gene name, if available, or by the protein/peptide product. Genes which may contain introns are distinguished from complementary DNA by appropriate phrases in the homograph data element. Finally, noninformational nucleic acids such as transfer RNAs, ribosomal RNAs, etc. are described by a small set of general controlled vocabulary terms and specific trivial designations.

CA INDEX NAME:

Deoxyribonucleic acid (Escherichia coli gene glgX)

CA INDEX NAME:

Deoxyribonucleic acid (Candida tropicalis clone pCC13
cytochrome P450 isoform 14DM gene)

CA INDEX NAME:

Deoxyribonucleic acid (Chlamydomonas reinhardti clone C321
cytochrome c messenger RNA-complementary)

CA INDEX NAME:

Ribonucleic acid (Escherichia coli glycine-specific transfer)

Figure 14. Nucleic Acid Homograph Names

PROBLEMS OF PROTEIN AND NUCLEIC ACID NOMENCLATURE

Some of the problems encountered in the nomenclature and representation of biopolymer sequences, and some of the solutions or temporary measures used by CAS are described below. The strength of the CAS Registry database is, obviously, that it is a chemical database. However, the chemical nature of the database does result in certain problems when applied to proteins and nucleic acids. An initial problem is the selection of the sequence structure for naming; the "chemical" nature of the CAS database demands a complete and unambiguous structure for registration and naming. The biological and experimental world of protein and nucleic acid sequences does not always easily fit this framework.

Incomplete or Partial Sequences

The problem with requiring complete and unambiguous information is that a lot of sequences that are reported are partial or incomplete and ambiguous but they are still of importance, such as sequences reported as a result of Edman Degradation (N-terminal sequence) *e.g.* A-B-C-D····X-Y-Z, or such as fragments resulting from partial hydrolysis, *e.g.* I-W-V-D-S L-E-W-F-K-P I-W-H-S M-C-S-A-L-R. When appropriate, incomplete or partial sequences can be named systematically or as specified length fragments of the "complete" sequence (Figure 15).

CA INDEX NAME:

1-142-Interferon γ (human lymphocyte protein moiety reduced), N²-(N-L-tyrosyl-L-cysteinyl)-

---

CA INDEX NAME:

Ferrate(4-), [L-valyl-L-glutaminyl-L-lysyl-L-cysteinyl-L-alanyl-L-glutaminyl-L-cysteinyl-L-histidyl-L-threonyl-L-valyl-L-glutamic acid cyclic (4→12′),(7→7′)-bis(sulfide) with 7,12-bis(1-mercaptoethyl)-3,8,13,17-tetramethyl-21H,23H-porphine-2,18-dipropanoato(6-)-N²¹,N²²,N²³,N²⁴]-, tetrahydrogen

CN Cytochrome c heme-undecapeptide (horse heart)

CN 11-Microperoxidase (horse heart cytochrome c heme-undecapeptide)
CN Heme undecapeptide

MF C84 H112 Fe N20 O21 S2 . 4 H

52 REFERENCES IN FILE CA (1967 TO DATE)

Figure 15. Incomplete or Partial Sequences

Flanking Regions

Nucleic acids present special problems. Even with modern sequencing technology, few genomes or other nucleic acid molecules are "completely" characterized. Generally, nucleic acid sequences presented in the literature are "fragments" of the entire genome which contain functional regions of interest (Figure 16). CAS certainly registers and names gene sequences identified within a much larger genome molecule -- but what is a gene?

```
 ...gacgctcacctagctgca ATG ACA GGT TCC ... GCT GGC CTG TAA ctgccgggtg...
<--Flanking region --> <--        Coding region        --> <-Flanking region->
```

Figure 16. Flanking Regions

Does a gene consist of a coding region alone, a coding region plus noncoding introns, the coding region plus the flanking regions on either side of the coding region which are transcribed to form messenger RNA but not translated to form protein, or even including further flanking regions which regulate the expression of the coding region but are not even transcribed into mRNA? For reasons of convenience, CAS has rather arbitrarily defined a gene as the coding region plus introns. Thus, the homograph nomenclature usually defines a nucleic acid fragment beginning at the initiator codon for protein translation and ending at the terminator codon.

Fusion Proteins

Another class of problems is the representation of defined structural variations which can be briefly illustrated by an example (Figure 17). Sequence-searchable databases of protein sequences allow the identification of the parental molecules involved in the construction of fusion proteins and facilitate naming by current methods. However, the number of possible constructs represents an ongoing challenge.

CA INDEX NAME:

18-86-Interferon γ, pre- (human lymphocyte clone pTG11 protein moiety reduced) 18-L-methionine-19-L-phenylalanine-20-L-tyrosine-21-L-phenylalanine-22-L-asparagine-23-L-lysine-24-L-proline-25-L-threonine-26-glycine-80-L-glutamic acid-81-L-valine-82-L-glutamic acid-83-L-valine-84-L-glutamic acid-85-L-phenylalanine-86-L-methionine-, **(86→ 1')-protein with** insulin-like growth factor I (human reduced)

Figure 17. Fusion Protein

A Mutant or Synthetic Gene Coding for a Substituted Protein

A somewhat similar problem for nucleic acids is the nomenclature for a mutant or synthetic gene coding for a substituted protein. Certainly, using substitution nomenclature for the nucleotide units would result in a bulky nomenclature term of limited if any utility. CAS represents such sequences by basing the name of the coded protein and using an abbreviated representation of the substituted amino acid residues within the protein (Figure 18). Deletions and insertions of residues can be named by similar procedures.

CA INDEX NAME:

Deoxyribonucleic acid, (human leukocyte clone pM21 interferon α2[Ala10Ala11Ala12Ala13Ala14Ala15]-specifying)

Figure 18. A Mutant or Synthetic Gene Coding for a Substituted Protein

```
RN  24800-07-5
CN  Insulin (ox-A reduced), 8-L-threonine-10-L-isoleucine-
    (9CI) (CA INDEX NAME)
CN  Insulin (dog-A reduced) (8CI)
CN  Insulin (human-A reduced)
CN  Insulin (rabbit-A reduced)
CN  Insulin (pig-A reduced)
CN  Insulin (sperm whale-A reduced)
CN  Porcine insulin A-chain (reduced)
MF  C99 H155 N25 O35 S4
STE 5:ALL,L
*** STRUCTURE DIAGRAM IS NOT AVAILABLE ***
40 REFERENCES IN FILE CA (1967 TO DATE)
```

```
RN  11061-68-0
CN  Insulin (ox), 8A-L-threonine-10A-L-isoleucine-30B-L-threonine-
    (9CI) (CA INDEX NAME)
CN  Insulin (human)
CN  3,4,44,45,90,91-Hexathia-8,11,14,17,20,23,26,29,32,35,38,
    41,48,51,54,57,60,63,66,69,72,75,78,81,84,86-hexacosaaza
    bicyclo[72.11.7]dononacontane, cyclic peptide deriv.  (9CI)
CN  Human insulin
CN  Insulin (Macaca fascicularis)
CN  Insulin (Macaca mulatta)
CN  Humulin R
CN  Novolin R
CN  Ultraphane
MF  C257 H383 N65 O77 S6
STE 5:ALL,L
*** STRUCTURE DIAGRAM IS NOT AVAILABLE ***
195 REFERENCES IN FILE CA (1967 TO DATE)
```

Figure 19. Insulins

Human Insulin Named as Analog of Insulin (ox)

For some polypeptides of particularly frequent and historically early interest, CAS has chosen a sequence from a single organism as the standard. Subsequent sequence structures with 50% or more similarity are named by substitution nomenclature of this standard. Probably, the best known examples of this nomenclature scheme are the insulins, where the so-called standard is bovine or ox insulin (Figure 19). Human insulin possesses a very similar sequence with only a substitution of threonine for alanine at positions 8A and 30B and a leucine for valine at position 10A. Human insulin is thus named as a substituted form of ox insulin. This results in the decidedly odd occurrence in the indexes of:
Insulin (ox), 23-leucine, of human, ...

System Limits

Proteins that are totally synthetic or of unidentified function, etc. might be named systematically. An amino acid sequence name for such a structure often exceeds current system limits (Figure 20).

CAS Registry Number: 86417-45-0

CA INDEX NAME:

**Adenosine,
[(dimethylethyl)benzoyl](tetrahydromethoxypyranyl)adenylyl-(3′→5′)-
(dimethylphenyl)(tetrahydromethoxypyranyl)uridylyl-(3′→5′)-
(dimethylphenyl)(tetrahydromethoxypyranyl)uridylyl-(3′→5′)-
[(dimethylethyl)benzoyl](tetrahydromethoxypyranyl)cytidylyl-(3′→5′)-
[(dimethylethyl)benzoyl](tetrahydromethoxypyranyl)cytidylyl-
(3′→5′)-[[(dimethylethyl)phenyl]acetyl](tetrahydromethoxypyranyl)
guanylyl-(3′→5′)-[[(dimethylethyl)phenyl]acetyl](tetrahydro
methoxypyranyl)guanylyl-(3′→5′)-[(dimethylethyl)benzoyl](tetra
hydromethoxypyranyl)adenylyl-(3′→5′)-[(dimethylethyl)benzoyl]
(tetrahydromethoxypyranyl)cytidylyl-(3′→5′)-(dimethylphenyl)
(tetrahydromethoxypyranyl)uridylyl-(3′→5′)-(dimethylphenyl)
(tetrahydromethoxypyranyl)uridylyl-(3′→5′)-[[(dimethylethyl)
phenyl]acetyl](tetrahydromethoxypyranyl)guanylyl-(3′→5′)-(dimethyl
phenyl)(tetrahydromethoxypyranyl)uridylyl-(3′→5′)-[(dimethyl
ethyl)benzoyl](tetrahydromethoxypyranyl)cytidylyl-(3′→5′)-
[(dimethylethyl)benzoyl](tetrahydromethoxypyranyl)cytidylyl-(3′→5′)-
[(dimethylethyl)benzoyl](tetrahydromethoxypyranyl)adenylyl-(3′→5′)-
[(dimethylethyl)benzoyl](tetrahydromethoxypyranyl)cytidylyl-(3′→5′)-
[(dimethylethyl)benzoyl](tetrahydromethoxypyranyl)cytidylyl-(3′→5′)-
[(dimethylethyl)benzoyl](methoxymethylene)-, octadecakis(2-chlorophenyl) ester,
5′-[2-(dibromomethyl)benzoate]**

Molecular Formula: $C_{601}H_{678}Br_2Cl_{18}N_{66}O_{184}P_{18}$

Figure 20. The Longest *CA* Index Name (1,578 characters)

Heterogeneity

A final class of problems in sequence nomenclature is the occurrence of heterogeneity. Biological sequences frequently have structural variability at defined positions which may or may not have any biological significance. Where no unique trivial designations are provided, CAS has found that strain or clone names provide a level of specificity that allows unique names to be provided for each unique sequence variation.

CONCLUSION

Both the current number of existing sequences and the rapid increase in the number of new sequences being reported require developing and maintaining nomenclature systems to keep up with advances in the field. Clear, unambiguous and unique names are needed in order to keep track of and to report biopolymer sequence information in abstracts, databases, and in indexes.

REFERENCES

1. IUPAC-IUB Joint Commission on Biochemical Nomenclature, "Nomenclature and Symbolism for Amino Acids and Peptides." *Eur. J. Biochem.* **1984**, 138(1), 9-37; *J. Biol. Chem.* **1985**, 260(1), 14-42; *Pure Appl. Chem.* **1984**, 56(5), 595-624.

2. IUPAC-IUB Commission on Biochemical Nomenclature, "Abbreviated Nomenclature of Synthetic Polypeptides (Polymerized Amino Acids)." *Eur. J. Biochem.* **1972**, 26, 301-304.

3. IUPAC-IUB Commission on Biochemical Nomenclature, "Abbreviations and Symbols for Nucleic Acids, Polynucleotides and their Constituents." *Eur. J. Biochem.* **1970**, 15, 203-208.

4. IUB Nomenclature Commission, "Nomenclature for Incompletely Specified Bases in Nucleic Acid Sequences." *Biochem. J.* **1985**, 229(2), 281-286; *Eur. J. Biochem.* **1985**, 150(1), 1-5; *J. Biol. Chem.* **1986**, 261(1), 13-17; *Mol. Biol. Evol.* **1986**, 3(2), 99-108; *Nucleic Acids Res.* **1985**, 13(9), 3021-3030; *Proc. Nat. Acad. Sci. USA* **1986**, 83(1), 4-8.

Development of Carbohydrate Nomenclature

Derek Horton

Department of Chemistry, The Ohio State University, Columbus, Ohio, U.S.A.

INTRODUCTION

The term "carbohydrate" was introduced early when gravimetric methods of analysis permitted the assignment of empirical formulas and it became recognized that starch, cane sugar, and the fibrous material of plant cell-walls were all comprised of carbon, hydrogen, and oxygen, with the latter elements being present in the same ratio as in water. The German term "Kohlenhydrate" and the French "hydrates de carbone" originated similarly. The generic suffix "-ose" in conjunction with the French "cellule" (for cell) led to the name "cellulose", the most abundant natural organic compound; however, the constitution of this compound as an acetal-linked linear biopolymer of a polyhydroxyaldehyde monomer was not recognized until some two centuries after its discovery.

EARLY TERMINOLOGY

As with many areas of natural product chemistry, source- or properties-derived trivial names were initially assigned, and with the recognition that "sugar" was only the first example among a series of other sweet-tasting, water-soluble carbohydrates, the name "glucose" was proposed for the sugar from grapes ("Traubenzucker"). With the introduction of the polarimeter, it was found that glucose was dextrorotatory, and was formed along with an isomeric, strongly levorotatory sugar when cane sugar ("saccharose", the English term "sucrose" was not proposed until much later) was subjected to mild acid hydrolysis to form the levorotatory mixture termed "invert sugar". Kekulé recommended the terms "dextrose" and "levulose", respectively, for glucose and the levorotatory isomer ("Fruchtzucker").

Emil Fischer's Contributions

Basic Sugar Structures: When Emil Fischer (1) commenced his monumental work of the sugars in the 1880's, he recognized that the sign of rotation was not a satisfactory basis for naming sugars and he returned to the name glucose, the name that has persisted since that time in all applications except perhaps among food technologists, where adherents of the old dextrose name are still to be found. In his 1890 address to the German Chemical Society, Fischer reviewed the known constitution of the isomeric $C_6H_{12}O_6$ sugars establishing that "Traubenzucker" ("grape sugar") is a 2,3,4,5,6-pentahydroxyhexanal, and that "Fruchtzucker" ("fruit sugar", levulose) is a 1,3,4,5,6-pentahydroxy-2-hexanone. He introduced the basis of classification of the sugar families that remains the foundation of current terminology: triose, tetrose, pentose, hexose, heptose, and so on according to the number of carbon atoms in the chain. He endorsed the

K. Loening (Editor)
The Terminology of Biotechnology:
A Multidisciplinary Problem

basic classification suggested by Armstrong that groups the sugars into aldoses and ketoses, and proposed that the "levulose" name be replaced by "fructose".

Chemical Correlations: From extensive chemical manipulations with great skill, coupled with brilliant reasoning, Fischer was able to correlate the families of sugars and perform a formal total synthesis of the natural (dextrorotatory) glucose from formaldehyde, along with its enantiomer, which he designated *l*-glucose, with "*d*-glucose" being the term identifying the natural enantiomer.

Families of sugars and their derivatives were grouped in series according to their chemical relationships in isomerization, chain-ascent, and chain-degradation reactions; German names corresponding to the derived aldonic acids and alditols were introduced, along with names for the 1,2-dicarbonyl sugars and their derived bis(phenylhydrazones) ("osazones"). The ready interconversion of "*d*-gluconic" and "*d*-mannonic" acids showed the close relationship of the two parent aldoses, and the chemical conversion of the pentose (+)-arabinose into "*l*-mannose" demonstrated this dextrorotatory pentose to belong to the family of "*l*-sugars", despite its dextrorotation. For the higher sugars obtained by synthesis, names were assigned according to the lower sugar precursor: mannoheptose, glucooctose, and the like.

Stereoisomers and the Fischer Projection: The van't Hoff—Le Bel concept of the asymmetrically substituted carbon atom and its consequences for the stereoisomerism among such molecules as the sugars provided the theoretical basis for establishing that an aldose having *n* asymmetrically substituted carbon atoms has 2^n stereoisomers. Fischer introduced his classic projection formulas for illustrating these relationships, leading to the familiar depiction of the 16 stereoisomeric hexoses, 8 pentoses, 4 tetroses, and 2 trioses. At the time, the absolute configurations could not be assigned, and an arbitrary selection (which later proved fortuitously to be correct in the absolute sense) had to be made in specifying that the triose (+)-glyceraldehyde is assumed to have the 2-hydroxyl group on the right in the projection formula. These formulas, as here exemplified for (+)-glucose and (-)-fructose, have since that time constituted the basis of reference for developing names for the sugars and their derivatives.

(+)-Glucose	(-)-Fructose
CHO	CH_2OH
|	|
HCOH	C=O
|	|
HOCH	HOCH
|	|
HCOH	HCOH
|	|
HCOH	HCOH
|	|
CH_2OH	CH_2OH

(+)-Glucose (-)-Fructose

Configurational Series: The possibility for confusion between the sign of optical rotation of a sugar and its stereochemical series was difficult to eliminate; it had to be emphasized that the formal relationship of the highest-numbered asymmetric carbon atom in a sugar to d- or l-glyceraldehyde is the determining factor. To minimize possible confusion, Rosanoff proposed replacement of the d- and l- terms by alternative symbolism that later was standardized as the current D and L terminology for specifying the configurational series.

EARLY 20TH CENTURY DEVELOPMENTS

By the turn of the century, the Fischer terminology and mode of depiction had come into widespread use. A system of fractional terminology proposed by Maquenne (2) [which would have used a symbolism of the type 2,4,5/3 to denote the stereoisomeric arrangement at C-2, 3, 4, and 5 of (+)-glucose] never gained popular support for the sugars, although it was incorporated into later terminology for the inositols, and might much later have facilitated the use of computerized data manipulation for stereoisomers.

Epimers: As in many fields of organic chemistry, compounds initially considered to be unique were later found to have isomeric or tautomeric modifications, as in the higher sugars synthesized by chain extension, where two new sugars arise from a single precursor. Thus two mannoheptoses are obtained by chain extension of natural mannose, and the prefixes α- and β- were used to differentiate the two isomers epimeric at the 2-position.

Cyclic Forms of the Sugars

The free sugars were found to exist, in most instances, as tautomeric mixtures of cyclic hemiacetals, designated α and β forms (later termed anomers), arising through reaction of one of the chain hydroxyl groups with the carbonyl group (3). The α anomer was defined as the one having the more-positive rotation for sugars in the D series (and the more-negative in the L series). The size of the ring was not definitively established in early work, but the most probable 5- or 6-membered ring structures were designated by the symbolism <1,4> or <1,5> following the name, or, as became general practice, by utilizing the infix "furano" or "pyrano" in the name. The Fischer formulas were adapted for depicting these cyclic forms, and the formal cis relationship between the anomeric hydroxyl group and the oxygen atom at the configurational carbon atom (C-5 in a hexose) in the α anomers (and trans relationship in the β anomers) became established as the basis for defining anomeric designators.

Haworth Projection and Conformational Formulas: The Fischer projection formulas, although convenient for defining the basic stereochemical relationships among the parent sugars, do not illustrate the true spatial geometry of the molecules, and the derived cyclic depictions, while correctly demonstrating connectivities and formal stereochemistry, give little impression of the real three-dimensional structure. Haworth introduced a new type of projection formula based on an arbitrary planar 5- or 6-membered ring, along with a conformational formula based, for the pyranose sugars, on the chair shape of the cyclohexane ring. Although chemical techniques of the time were insufficiently advanced for the conformational formulas to be properly appreciated, they provide a close approximation to the real shape of pyranoid sugars and later became the most widely used and realistic structural representations for the sugars.

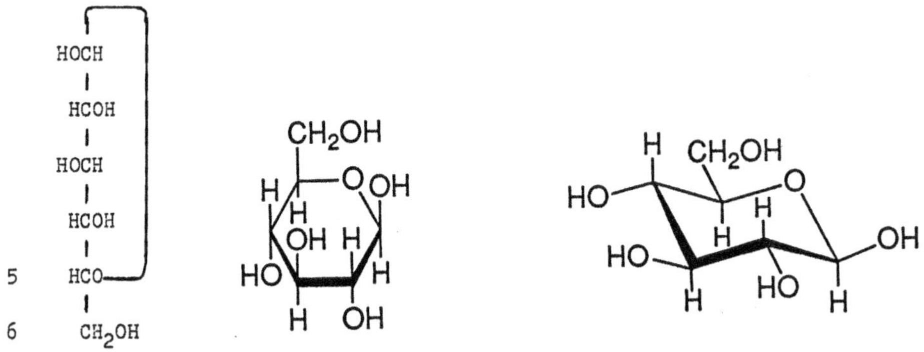

(Fischer) (Haworth projection) (Haworth conformation)

Cyclic Structural Depictions for β-D-Glucopyranose

<u>Deoxy Sugars</u>: Aldohexoses having the terminal hydroxymethyl group replaced by methyl were initially termed "methylpentoses", but recognition that deoxygenation of a parent sugar may be effected at other positions in the molecule led to such terms as "2-desoxyglucose" (alternatively termed "2-desoxymannose") in which a locant specifies the point of deoxygenation; the duplication of names resulting from deoxygenation at an asymmetric center was later eliminated with the adoption of systematic configurational designators. The archaic form persists with the trivial name "2-deoxy-ribose" ("2-deoxyarabinose").

<u>Glycosides</u>: The recognition that glycosides, formed by acid-catalyzed reaction of a sugar with an alcohol or phenol, are mixed full acetals having the anomeric hydroxyl group replaced by the RO group of the alcohol or phenol, enabled terminology and symbolism to follow closely that used for the parent cyclic sugars. The name "α-methylglucoside" was subsequently formalized as methyl α-D-glucopyranoside, following general practice for nomenclature of acetals. The term "aglycon" came into use for the substituent at the anomeric center. The generic term "glycosyl" for the group derived from the cyclic sugar by removal of the anomeric hydroxyl group permitted great flexibility in naming complex glycosides.

<u>Oligosaccharides and Polysaccharides</u>: When the aglycon in a glycoside is another sugar, the resultant product is a disaccharide, glycosylation of which results in a trisaccharide, and so on. The general term "oligosaccharide" was introduced for such carbohydrates still of relatively low molecular weight, while the term "polysaccharide" was adopted for large molecules of similar constitution but having macromolecular properties. Cellulose could thus be formulated as a linear high polymer of β-D-glucopyranose units joined in glycosidic union at position 4:

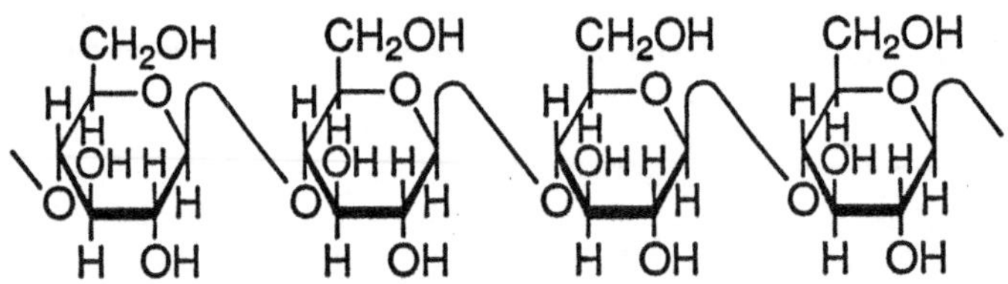

The dividing line between oligosaccharide and polysaccharide has remained somewhat arbitrary, but the upper limit for classification as an oligosaccharide has tended to increase as methods for precise structural attribution have advanced. The current trend is to employ the former term when a specific structure can be defined, and the latter when a polydisperse macromolecular structure is involved having an undefined large number of monosaccharide residues.

CARBOHYDRATE NOMENCLATURE IN THE 1930's

The International Union of Chemistry did not offer directives for the terminology of carbohydrates during the development of the Geneva system for organic nomenclature, nor at the Liège Council in 1930. The active development of carbohydrate chemistry during the first four decades of the 20th century without a developing basis of formalized terminology led to a profusion of trivial names that were inexact and did not indicate precise structure. Such names as "chitosamine", "diacetoneglucose", "acetobromomannose," "α-methyltetramethylglucose", and many others were only understandable in specific structural terms by specialists in the field.

ROLE OF NATIONAL CHEMICAL SOCIETIES

American Chemical Society

In an effort to keep a rational base of understandable terminology in this rapidly advancing field, a committee of the American Chemical Society, including Horace S. Isbell, R. Stuart Tipson, and Charles D. Hurd began work in 1939 on preparing reports to be considered by the ACS Committee on Nomenclature (4). In a series of reports during the next several years, the outline of modern terminology was delineated: the numbering of the sugar chain, locants for substituent prefixes, use of D and L, and italicized prefixes (such as glycero, erythro, ribo, allo) for designating stereochemistry at one, two, three, or four configurational centers along a sugar chain. Combinations of such prefixes were specified for the higher-carbon sugars, to replace the confusing trivial names in the earlier literature. A committee of the ACS Division of Carbohydrate Chemistry led by Melville L. Wolfrom prepared the final version of the report, which was approved by the ACS Council in 1948 (5).

British—American Rules: Not all questions were resolved in the 1948 ACS report; agreement could not be reached on many aspects of terminology for unsaturated sugars, branched-chain sugars, and oligo- and poly-saccharides, and differences still persisted between British and American usage, as well as with the older German terminology. A joint British—American committee revised the 1948 report, resolving many differences between British and American usage and published in 1952 its "Rules of Carbohydrate Nomenclature" (6). The established parent sugar names were retained, but more-rational names were recommended to replace many older trivial names, for instance 2-amino-2-deoxy-D-galactose instead of galactosamine or chondrosamine, and the O- prefix was specified for substituent groups to designate substitution on oxygen rather than carbon. Redundancies in the naming of deoxy sugars were eliminated by specifying the use of a parent name, for example 3-deoxyhexose, in conjunction with an italicized configurational designator, for example D-ribo, that specifies only the sequence of stereocenters actually present.

Continuing work by the joint British—American Committee led to an expansion and revision of the Rules, which were published (7) in 1963 with the approval of the American Chemical Society and the Chemical Society of Great Britain. Again, areas where uniform agreement could not be reached were omitted, but the document provided a basis for specifying a unique name for a very large proportion of the carbohydrate derivatives known, and further eliminations of trivial names were proposed. The anomeric symbolism α and β was specified in relation to a "reference atom" that is the highest-numbered asymmetric carbon atom in the chain for the common sugars. For the higher-carbon sugars for which a multiple configurational prefix is required, the reference atom relates to the configurational prefix closest to the anomeric center, for example methyl L-*erythro*-α-D-*galacto*-octopyranoside.

Although the 1963 report did not provide formal specifications for the terminology of polysaccharides, it set out a basis for symbolic depiction of oligo- and poly-saccharide structure by defining abbreviations for the parent sugars and their ring sizes, for example Gal (galactose), Glc (glucose), Fru*f* (fructofuranose), and GulA (guluronic acid), together with linear or polydirectional symbolism for structures of complex carbohydrates, as for instance: β-D-Glc*p*-(1→4)-(β-D-Glc*p*)$_n$-D-Glc (cellulose). This terminology provided a framework for the later development of computerized databases for structures of complex carbohydrates.

IUPAC-SPONSORED CARBOHYDRATE NOMENCLATURE

General Recommendations

In 1964 the Commission on the Nomenclature of Organic Chemistry of the International Union of Pure and Applied Chemistry (IUPAC) appointed a committee to consider the Anglo-American rules and prepare a set of IUPAC Rules for Carbohydrate Nomenclature. The work of this committee led to the publication (8) in 1971 of Part 1 of the Tentative Rules for Carbohydrate Nomenclature as a joint report of IUPAC and the International Union of Biochemistry (IUB). The format of these Rules largely followed that of the Anglo-American document, but again deferred for later consideration such areas as unsaturated and branched-chain sugars, conformational specifications, and polysaccharide nomenclature.

Trivial and Systematic Names: The 1971 IUPAC Rules further rationalized the parent names of the sugars and the use of italicized configurational designators by specifying systematic names for all parent sugars, thus: D-*gluco*-hexose, L-*arabino*-pentose, D-*arabino*-2-hexulose, respectively, for the trivial names D-glucose, L-arabinose, and D-fructose. The trivial names were specified as the preferred forms for common usage, with the systematic ones to be employed in constructing names of highly modified structures.

Configurational Prefixes and the Reference Atom: Agreement could not be reached on all areas in the 1971 document, notably concerning the use of configurational prefixes in deoxy and dicarbonyl sugars, the specification of the reference atom in defining anomeric symbols, and in the retention of older trivial names. Multiple configurational prefixes were recommended, for instance 3-deoxy-D-*erythro*-L-*glycero*-hexose, with the "Anglo-American" usage 3-deoxy-D-*arabino*-hexose being given as a permissible alternative. The multiple-prefix version has not enjoyed wide acceptance over the simpler, single-prefix version. The reference atom was redefined as the highest-numbered asymmetric carbon atom, which caused the anomeric symbolism to be reversed for some of the higher-carbon sugars, thus the older form methyl L-*erythro*-α-D-*galacto*-octopyranoside became methyl β-L-*erythro*-D-*galacto*-octopyranoside. Although there is no fundamental ambiguity between the two specifications when used with the full names, confusion is possible when abbreviated symbolism is used for the higher-carbon sugars in conjunction with anomeric designators.

Subsequent Recommendations

The IUPAC-IUB committee continued its work on those areas not covered by the 1971 report and in 1980 issued separate published documents with recommendations for conformational nomenclature for pyranoid and furanoid sugars (9), branched-chain monosaccharides (10), unsaturated monosaccharides (11), abbreviated terminology of oligosaccharide chains (12), and polysaccharide nomenclature (13). These recommendations finally provided an internationally agreed basis for terminology where hitherto there had been considerable confusion. The older Reeves conformational designators *C1* and *1C* became $^4C_1$ and $^1C_4$ for the common pyranoid chair conformations. The branched-chain terminology addressed the question of configurational designators by using the Sequence Rule priorities at the branch point and then allocating conventional stereochemical prefixes for the sugars; although not explicit in the document, this principle may also be used to advantage in naming isotopically modified sugars.

Unsaturated Sugars: The long vexing problem of naming unsaturated sugars, which started when Emil Fischer introduced the unfortunate term "glycal", was resolved (11) by the conceptually simple expedient of naming the saturated parent structure and inserting the infix "en" with an appropriate locant. Although "D-glucal" by this system became the more cumbersome 1,5-anhydro-2-deoxy-D-*arabino*-hex-1-enitol, the latter name is readily interpreted as a specific structure by the established sugar terminology.

Oligosaccharide and Polysaccharide Symbolism and Terminology: The 1980 oligosaccharide chain terminology extended the three-letter symbolism and polydirectional arrays of the 1963 report to include specifications for aminodeoxy sugars and other types of substituted sugars, and provided an alternative condensed symbolism for use when maximum compression of structural information is desired. The codification of polysaccharide terminology brought long-sought guidelines in an area where accords had been difficult to reach because of disagreements between workers in the field and with specialists in other areas of macromolecular chemistry. The generic term "glycan" adopted for polysaccharides is not quite synonymous with poly(glycose), as it is restricted to molecules in which the monomeric residues are joined by glycosidic linkages. Although changes in such long-established names as starch, cellulose, and glycogen were not advocated, the "-an" suffix was recommended for newly discovered polysaccharides as well as for polysaccharides whose earlier trivial names had not become firmly entrenched.

Conformation of Polysaccharide Chains: Specifications for conformational symbolism in polysaccharide chains were also developed by an IUPAC-IUB committee and were issued (14) in 1981. These notations follow the general principles of macromolecular conformational symbolism used in the polypeptide and polynucleotide fields, and specify the torsional angles from one glycosyl residue to the next monosaccharide of the chain, along the bonds connecting the two residues.

CURRENT AND FUTURE TRENDS

New Areas of Carbohydrate Nomenclature: Although no additional formal IUPAC-IUB documents on carbohydrates have been issued since 1981, much further work has been done through working drafts to revise the 1971 Tentative Recommendations. Addressed are such outstanding questions as terminology for sugar molecules fused to heterocyclic rings, the provision of alternatives to the inositol nomenclature for cyclic sugar analogs in which the ring oxygen atom has been replaced by carbon, and extensions of the glycosyl terminology to permit radical names for the sugars to be specified from positions other than the anomeric carbon atom. Working drafts of documents on these aspects have been used as an ongoing basis for current nomenclature practice, pending formal ratification of final documents. A revision of the 1971 document seeks to establish uniform recommendations by eliminating such redundancies as duplicate systems of configurational prefixes in favor of a single configurational designator wherever possible, by defining explicitly the choice of the reference atom in specifying the anomeric designators, and by expanding the number of illustrative examples, with particular selection from active current areas of research.

Enzymes Acting on Carbohydrates: The terminology for enzymes acting on carbohydrates includes the formal names of the substrate molecules, and some variations exist between the names used in the current Enzyme Nomenclature (15) and the IUPAC-IUB carbohydrate recommendations; these are scheduled to brought into uniformity in a forthcoming edition of the enzyme report.

Electronic Data Retrieval

A problem of rapidly increasing significance in chemistry is the electronic handling and retrieval of structural chemical data. In the carbohydrate field, the wide stereochemical diversity within a single parent framework of a monosaccharide and its derivatives complicates the access of information by established computer-search procedures. The possibilities for automated search become rapidly exhausted as the molecular complexity increases, necessitating new approaches for the electronic handling of information on oligo- and poly-saccharides.

Computer Database for Complex Carbohydrate Structures: A structural database for complex carbohydrates (CCSD) that can be searched by a computer program, CARBBANK, has been developed under the auspices of an international group (16), and has the objective of encoding structural information on all known oligosaccharides and polysaccharide sequences in a format that permits information searches from a variety of approaches.

Information encoded in the CCSD database utilizes the three-letter abbreviations as codes for the monosaccharide sugars, with complementary modifiers for ring size, anomeric and enantiomeric designation, and structural modification (uronic acids, aminodeoxy sugars, and so on), in a format based on the recommendations in the 1980 IUPAC-IUB oligosaccharide document. The CARBBANK program permits searches for whole structures, substructures, or text, and provides computer access to a large body of literature that cannot be searched electronically by other database methodology. Its conceptual basis is related to the polynucleotide database GENBANK and the structural database for polypeptides and proteins, but the multiplicity of linkage arrangements, monosaccharide types, and the possibilities for branching add complexity to the encoding and handling of carbohydrate structures.

Future Development of Carbohydrate Databases: Ongoing challenges for this type of approach, for which consistent and unambiguous nomenclature is required, include computerized methods for generating three-dimensional structures from encoded symbolic structures, correlation with crystallographic and NMR data, and the probing of molecular host—guest interactions.

REFERENCES

1 Fischer E (1890) Ber 23: 2114
2 Maquenne L (1900) Les sucres et leurs principaux dérivés, Gauthier-Villars, Paris
3 Elsner H (1935) Tollens-Elsner's Kurzes Handbuch der Kohlenhydrate, 4th edition, Johann Ambrosius Barth, Leipzig
4 Hurd CD (1989) J Chem Ed 66:984
5 Rules of carbohydrate nomenclature (1948) Chem Eng News 26:1623
6 Rules of carbohydrate nomenclature (1952) J Chem Soc 5108; Chem Eng News 31:1776 (1956).
7 Rules of carbohydrate nomenclature (1963) J Org Chem 28:281
8 IUPAC-IUB Tentative rules for carbohydrate nomenclature: Part 1, 1969 (1971) Biochemistry 10: 3983 correction 4995; Eur J Biochem 21: 455
9 Conformational nomenclature for five and six-membered ring forms of monosaccharides and their derivatives: recommendations 1980 (1981) Arch Biochem. Biophys 207:469; Eur J Biochem 111:295; Pure Appl Chem 53:1901
10 Nomenclature of branched-chain monosaccharides: recommendations 1980 (1981) Eur J Biochem 119:5 corrections 125:1; Pure Appl Chem 54:211 (1982)
11 Nomenclature of unsaturated monosaccharides: recommendations 1980 (1981) Eur J Biochem 119:1 corrections 125:1 (1982); Pure Appl Chem 54:207 (1982)
12 Abbreviated terminology of oligosaccharide chains: recommendations 1980 (1982) Eur J Biochem 126:433; J Biol Chem 257:3347
13 Polysaccharide nomenclature: recommendations 1980 (1982) Eur J Biochem 126:439; J Biol Chem 257:3352; Pure Appl Chem 54:1523
14 Symbols for specifying the conformation of polysaccharide chains: recommendations 1981 (1983) Eur J Biochem 131:5: Pure Appl Chem 55:1269
15 IUB, Enzyme nomenclature (1984) Academic Press, Orlando
16 CARBBANK and complex carbohydrate structure database (1989) University of Georgia Complex Carbohydrate Center

Enzyme Nomenclature

Edwin C. Webb, F.R.A.C.I.

Department of Chemistry and Biochemistry,
James Cook University of North Queensland,
Townsville, Queensland 4811, Australia

HISTORY

For many years there has been an almost continuous series of standing committees, organized by the International Union of Biochemistry or the International Union of Pure and Applied Chemistry, or both, dealing with nomenclature problems specific to biochemists. The present arrangement, described by Kurt Loening earlier in this symposium, is a Nomenclature Committee of I.U.B., and a Joint IUPAC/IUB Committee on Biochemical Nomenclature; the Committees have overlapping membership and normally meet in combined sessions. For most of the time such committees have existed, and certainly since the late 1950's when I.U.B. set up the original Enzyme Commission under the Chairmanship of Malcolm Dixon, enzyme nomenclature has occupied more of their time than any other section of biochemical nomenclature. In this paper I hope to show why this is so.

Those of you who are chemists unfamiliar with biological systems may wonder why there should be any special nomenclature for enzymes. Enzymes are, almost entirely, proteins, and indeed this has been made part of the definition of enzymes. To quote a 1958 textbook on enzymes (1): "*An enzyme is a protein with catalytic properties due to its power of specific activation.*" This may need a little expansion to include other biological polymers like nucleic acids, but the basic principle still holds: an enzyme is a unique chemical molecule which, in theory, could be named by the normal rules of organic chemistry. There would be difficulties about doing this because of the size and complexity of the molecules, but these are the common difficulties about biological substances which have frequently led biochemists to invent trivial nomenclatures to replace impossibly complex systematic chemical names. Let me illustrate this question briefly by considering its application to proteins.

K. Loening (Editor)
The Terminology of Biotechnology:
A Multidisciplinary Problem

There are two aspects to a nomenclature system. It must provide a naming system which identifies a particular compound: this could be a code or even simply a number on an agreed list. Secondly, the name may give some information about the nature of the compound, and of course a full systematic nomenclature should give full and unambiguous information. Fortunately for biochemists there are often many common features in groups of biological compounds, and these features can be coded or, as it were, "taken as read". Thus proteins are built up of 20 or so building blocks which are a very disparate group of compounds, but they are all based on an α-aminocarboxylic acid structure with an L-configuration at the α-carbon, and are linked together by amide bonds, the α-peptide links. All proteins can therefore be represented as one or more peptide chains, as in Figure 1.

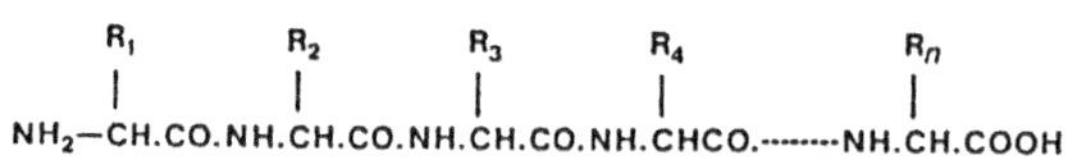

Figure 1. A PEPTIDE CHAIN

In 1974 a standard set of 3-letter abbreviations for the amino acids themselves, which could be used to define peptide structures, was put forward jointly by the Commission for Nomenclature of Organic Chemistry and the Commission on Biochemical Nomenclature.

The backbone structure shown in the slide is coded by dashes between the coded amino acids, as in Figure 2.

Alanyl-arginyl-cysteinyl-glycyl-histidyl-leucyl-serine

Ala-Arg-Cys-Gly-His-Leu-Ser

Figure 2. TRIVIAL NOMENCLATURE FOR A PEPTIDE

A good deal of information is defined by this representation. The free amino group is assumed to be at the left end of the chain, the free carboxyl at the right; the hyphens represent -CO.NH- bonds; all amino acids shown are of the L-configuration at the α-carbon, unless the 3-letter symbols are preceded by a small capital D. Since 1974 the symbolism has been extended to cover synthetic modifications of amino acids and bonds other than α-peptide bonds, and some group symbols, *e.g.* Asx means either aspartic acid or asparagine.

The complete amino acid sequence of many purified enzymes is now known, and can be printed reasonably satisfactorily by this method; many enzymes are included in the published *Atlases of Protein Sequences and Structures*. Such representations are hardly useful as names, and give little help in understanding the mode of action of enzymes. The formation of the catalytic environment depends on the bringing together by the tertiary structure of the protein of active radicals, so that a complete and useful description of an enzyme as a substance must include a 3-dimensional crystal structure.

CLASSIFICATION OF ENZYME-CATALYSED REACTIONS

All this is, in any case, a digression, because enzyme nomenclature has been directed to a classification and description of their catalytic action, which is, after all, the primary interest, rather than to their chemical structure. The chronology of enzyme research is usually that a catalytic reaction is detected in a tissue extract or some similar crude biological preparation; later the enzyme is extracted and purified, and still later work may be done on the protein chemistry of the enzyme and its mode of action. Inevitably they are at first named in terms of a more or less specific chemical reaction which they catalyse.

It was Duclaux in 1898 who first suggested a system which in essence has persisted to the present day, that of adding ***-ase*** to the chemical name of a substrate to indicate that the substance is acted on by the enzyme. Without any other addition, the catalysed reaction is hydrolysis; thus *glucose-6-phosphatase* acts on glucose-6-phosphate, hydrolysing it to glucose and phosphate. For reactions other than hydrolysis, a prefix indicating the kind of reaction catalysed is attached to the *-ase; e.g. decarboxylase, phosphotransferase*, which are self-explanatory names. Enzyme names are of the type ***substrate function-ase***; where two substrates are involved, they are separated by a long dash or a colon. For historical reasons, a few of the proteolytic enzymes end in *-in*, such as *pepsin* or *chymotrypsin*. All other enzyme names in current use end in *-ase*.

The original I.U.B. Enzyme Commission, which reported in 1961 (2), attempted to rationalize enzyme names by using a classification system which was then picked up in the functional prefix of the name. Their classification system, based on six main classes, has continued to be the basis of enzyme classification and nomenclature, although a number of new subclasses has been added. The classification is shown in Figure 3.

Class 1 includes all enzymes dealing with oxidation, by far the largest class. Class 2 includes *acyltransferases, phosphotransferases, aminotransferases* and so on. Class 3 includes *esterases, phosphatases, glycosidases, etc.* Class 4 reactions can be represented as addition of a molecule across a double bond; uniquely, two substrates are involved in one direction but only one in the other; *e.g. decarboxylases, dehydratases, aldolases.* Class 5 is self-explanatory; class 6 is the important group by means of which synthetic reactions are coupled with breakdown of nucleoside triphosphate.

GROUP	NAME	GENERALISED REACTION
1.	Oxidoreductases	$AH_2 + B = A + BH_2$ *or* $AH_2 + B^+ = A + BH + H^+$
2.	Transferases	$AX + B = A + BX$
3.	Hydrolases	$A{-}B + H_2O \rightarrow AH + BOH$
4.	Lyases	$A{=}B + X{-}Y = \underset{X\ \ Y}{A{-}B}$
5.	Isomerases	$A = B$
6.	Ligases	$A + B + NTP \rightarrow A{-}B + NDP + P$ *or* $A + B + NTP \rightarrow A{-}B + NMP + PP$

Figure 3. CLASSIFICATION OF ENZYMES

Perhaps the most important action of the Enzyme Commission was the invention of a numbering system (2), which has been widely used and continues to be used. Four numbers are separated by dots, and the whole number is preceded by EC, originally meaning Enzyme Commission. Figure 4 shows a typical example.

The first number identifies one of the 6 major classes (oxidoreductases), the second and third indicate sub-classes (defined differently in each class), and the last is simply a numerical listing in an agreed standard list.

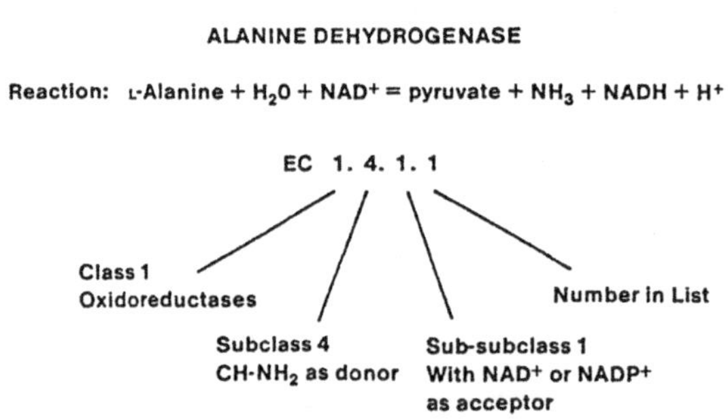

Figure 4. ANATOMY OF AN EC NUMBER

The 1961 Enzyme Commission report contained an enzyme list with 712 entries. In subsequent editions the number went to 875 in 1964, 1770 in 1972, 2122 in 1978, and 2477 in 1984 (3). Some of the earlier entries have been altered or deleted as a result of more recent work; to avoid ambiguities, the numbers are not reused, and are listed with cross-references. Two supplements (4,5) have appeared in the *European Journal of Biochemistry* since the publication of *Enzyme Nomenclature (1984)* (3), and a third is in press. The total number of 'live' entries is now 2961, and in addition there are 257 cross-references for deleted or transferred entries.

PRESENT PROBLEMS

Updating of the I.U.B. Enzyme List is an ongoing process. New entries arise partly from culling of the literature by members of NC-IUB, but more importantly by authors writing to Edwin Webb in Townsville or to NC-IUB drawing attention to their own work. Some journal editors have also drawn attention to enzymes not in the enzyme list; the European journals (*Biochimica et Biophysica Acta* and the *European Journal of Biochemistry*) have been particularly helpful. Through the good offices of Kurt Loening, *Chemical Abstracts* has helped by producing an annual list of enzyme names which appear to be missing from the Enzyme List.

If an enzyme name is to define a catalysed reaction, the same type of problem that I discussed just now about peptide structures comes up again. Substrates are often very complex, and common features are often omitted from the enzyme name. Thus *X dehydrogenase* normally implies that either NAD or NADP is the hydrogen acceptor for the oxidation of X; *A—B ligase* implies that A and B react to produce AB with a simultaneous hydrolysis of ATP. These short-cuts or coding apply to the listed **Recommended Name** and where it exists, the trivial biochemical nomenclature for the substrates is used in these names. The reaction is spelt out more unambiguously in the listed **Reaction**, and also in the **Systematic Name**, based on this reaction. More or less complete systematic chemical nomenclature is used for these entries. The systematic name is not greatly used, since the EC number is a code which, by reference to the Enzyme List, defines the enzyme fairly precisely.

CMP-*N*-acetylneuraminate—(α-*N*-acetylneuraminyl-2,3-β-galactosyl-1,3)-*N*-acetylgalactosaminide α-2,6-sialyltransferase

Figure 5. RECOMMENDED NAME FOR EC 2.4.99.7!

I said earlier that one purpose of nomenclature is to provide a name which clearly identifies a substance. To be useful such a name should be short enough to use in print, and euphonious enough to use in spoken communication. Despite the contractions I have been talking about, I would not claim that these objectives have been always achieved. Figure 5 shows a recent Recommended Name from the published list. Any attempt to read it will show that this is not a name which will be readily used!

I have been describing a system of classification and nomenclature, based on the catalysed reaction, which has worked well for nearly 30 years. Problems do arise in applying the system, and I have only time to summarise them briefly. Many enzymes catalyse more than one reaction, in different classes, and it then becomes an arbitrary choice as to where they should be listed. Many enzymes catalysing hydrolytic reactions will also, under suitable conditions, catalyse transfer reactions; they are normally listed as hydrolases. The transaminases are so listed, even though the reaction of an amino acid and a keto acid is also an oxidoreduction of both molecules.

Special difficulties arise when several catalytic proteins are involved in a complex, as in the mitochondrial respiratory chain, or the pyruvate dehydrogenase complex. How far should the separate proteins be listed as separate enzymes, and what reaction is then to be specified? Should the complete functional complex be listed as well?

Many entries in the list really describe a group of enzymes of overlapping specificity, and it becomes a matter of subjective judgement as to when new work describes an enzyme which is sufficiently different to justify a separate entry and EC number.

LONG-TERM PROBLEMS

I want to go on in the last part of this paper to discuss problems that are arising with the whole system.

This system, as I have said, is based on the proposition that the catalysed reaction is often all that is known about an enzyme. Increasingly, we know a great deal more, particularly in three areas. Often we know about the structural chemistry of the enzyme, especially the presence of non-protein prosthetic groups. Secondly, we may have information about the genetic relationships of the enzyme protein. Thirdly, in many cases we have a quite precise knowledge of the catalytic mechanism.

Biochemists working with such enzymes tend to argue for a new classification based on enzyme structure or mechanism. As far back as 1974 Paul Engel (6) wrote:

> "As I understand it, the EC classification system at present is based entirely upon a functional distinction between enzymes in terms of reaction stoichiometry and substrate specificity. At one time that was the only obvious and useful criterion for distinguishing enzymes and for many enzymes it remains so. However, for an ever-growing number of enzymes we have two other sorts of information which threaten to make EC classifications appear arbitrary, obsolete and confusing. First of all there is comparative information regarding function properties of enzymes from many species. This often reveals the distinct entities of the EC manual to be part of a continuum......
>
> The second major type of information which now challenges the basis of existing classification, is structural information. Initially this is molecular weight information, and perhaps this is insufficient, but for many enzymes we also have amino acid sequences. To give you a specific example again, there ought to be something in the classification to indicate the difference between zinc-dependent alcohol dehydrogenases of molecular weight *c.* 38,000 (*e.g.* horse and yeast ADH*) and zinc-independent ADH's of subunit molecular weight *c.* 25,000 (*e.g. Drosophila* ADH). Preferably this should also indicate the close homologies that exist between ADH's of both classes and separately classified sugar alcohol dehydrogenases, again Zn-dependent and Zn-independent. ..."

Unfortunately, we have already adopted these other principles in a partial and unthinking way. For example, EC 1.4.3.4 and 1.4.3.6 have recommended names of ***amine oxidase (flavin-containing)*** and ***amine oxidase (copper-containing)*** respectively. But the problems are particularly obvious in sub-class 3.4, the peptide hydrolases. The sub-sub-classes are based on very mixed criteria. 3.4.17 and 3.4.24 are the *metallo-carboxypeptidases* and the *metallo-proteinases*, in other words, are based on the non-protein chemistry of the enzymes. On the other hand, 3.4.16, *serine carboxypeptidases*, 3.4.18, *cysteine carboxypeptidases*, 3.4.21,

*Incidentally, the Nomenclature Committee of I.U.B. does not encourage the use of abbreviations for enzyme names. With around three thousand known enzymes, the possibility of ambiguity arising is very great.

serine proteinases, 3.4.22, *cysteine proteinases*, and 3.4.23, *aspartic proteinases*, define the nature of the active centre and the catalytic mechanism. The other sub-classes largely define the nature of the substrate, *e.g.* 3.4.13, *dipeptide hydrolases*, or 3.4.14, *dipeptidylpeptide hydrolases*.

Some examples will show the results of this mixed approach. Figure 6 is taken from Enzyme Nomenclature (1984) (3).

EC 3.4.16.1	Serine carboxypeptidase
EC 3.4.17.4	Glycine carboxypeptidase
EC 3.4.21.26	Prolyl endopeptidase
EC 3.4.13.12	Methionyl dipeptidase
EC 3.4.23.11	Thyroid aspartic proteinase

Figure 6. AMBIGUOUS NAMES FROM ENZYME NOMENCLATURE (1984)

These entries look quite similar. However, the first name shows that serine is in the active centre of the enzyme protein; the next that the enzyme acts on peptides with a C-terminal glycine; the third hydrolyses prolyl bonds; and the fourth hydrolyses methionyl-dipeptides. However, in the last example we revert to mechanism, showing that an aspartic residue is in the active centre. Thyroid indicates the source of the enzyme, which is not usually used as a basis for nomenclature or classification.

The Nomenclature Committee of I.U.B. is aware of the inconsistencies, and has attempted to improve the classification of this sub-class. Nevertheless Figure 7 shows some examples of the continuing problem, from the Supplements published since 1984.

EC 3.4.21.55	*Bitis gabonica* serine peptidase
EC 3.4.24.11	Membrane metalloendopeptidase
EC 3.4.21.58	Prohormone serine proteinase
EC 3.4.22.21	Yeast cysteine proteinase E
EC 3.4.11.16	Aminopeptidase X-Trp
EC 3.4.11.17	Tryptophan aminopeptidase
EC 3.4.17.3	Lysine(arginine) carboxypeptidase

Figure 7. SOME RECENT ADDITIONS TO THE ENZYME LIST (SUPPLEMENTS 1 AND 2)

Looking at those above the dotted line, the names indicate both source and mechanism; those below relate only to the catalysed reaction, showing the amino acid residue actually removed from the substrate. Yet note the similarity of form between numbers 3, 4, 6 and 7!

There are some short-term solutions to deal with the more obvious ambiguities. The pattern was set when EC 3.4.16.1 was named ***serine carboxypeptidase***. In retrospect, this was an unfortunate decision. Application of the principles applied to the rest of the enzyme list, which I have been describing, suggests an enzyme hydrolysing peptides at the C-terminal position, either removing a serine residue, or alternatively only acting if there is a serine residue close to the terminal peptide bond. (These two interpretations are not distinguished by names in the present enzyme list.) Suggestions under consideration by a working party convened by Alan Barrett are to rename this enzyme (EC 3.4.16.1) ***serine-type carboxypeptidase*** or ***serine—dependent carboxypeptidase***, to indicate a serine residue at the active centre of the enzyme.

However, this discussion raises a broader question: has the time come to devise a new system of classification and nomenclature of enzymes based on catalytic mechanism or genetic relationship or both?

My own view, for what it is worth, is that although a great deal of such information is available, it still covers only a small part of the more than two thousand known enzymes. It may be sensible to develop alternative classifications for small groups of enzymes, but I believe that for the time being there is considerable value in continuing the production of an agreed list of enzymes with code numbers and names based on the type reaction catalysed by each enzyme. Certainly this seems to be the view of the International Union of Biochemistry, which has agreed to publish a new edition of *Enzyme Nomenclature: Recommendations of the Nomenclature Committee of the International Union of Biochemistry*. Hopefully, this will be on the market late in 1991, together with a machine-readable version.

REFERENCES

1 Dixon M, Webb EC (1958) Enzymes, p.5

2 Commission on enzymes of the International Union of Biochemistry (1961) Report. Pergamon Press Oxford

3 International Union of Biochemistry (1984) Recommendations of the Nomenclature Committee of the International Union of Biochemistry on the nomenclature and classsification of enzyme-catalysed reactions. Academic Press, New York

4 International Union of Biochemistry (1986) Enzyme Nomenclature: Recommendations 1984: Supplement 1: Corrections and additions. Eur J Biochem 157:1

5 International Union of Biochemistry (1989) Enzyme Nomenclature: Recommendations 1984: Supplement 2: Corrections and additions. Eur J Biochem 179:489

Engel P (1974) Letter to the Nomenclature Committee of I.U.B.

Interferon Nomenclature

M.I. Johnston, R. J. Black and M.W. Myers

Division of AIDS, National Institute of
Allergy and Infectious Diseases
National Institutes of Health, Bethesda, MD 20892

Interferons were discovered in 1957 and subsequently divided into antigenically distinct types: type I, represented by the leukocyte and fibroblast types (generally acid stable), and type II, represented by immune interferon (generally acid labile). With cloning of the interferon genes and formation of an ad hoc interferon nomenclature committee, the names interferon-α, interferon-β, and interferon-γ to refer to the pure species that comprise the predominant class of leukocyte, fibroblast and immune interferons, respectively, were proposed and met with general acceptance among the scientific community. The evolution of the current nomenclature system for interferon proteins, recent studies on interferon-β2, acceptance of the term IL-6, and unresolved nomenclature issues in the interferon field are overviewed here.

NOMENCLATURE EVOLUTION

Pre-1980

In 1957, the name interferon (IFN) was given by Isaacs and Lindemann to an inducible protein factor produced by chicken chorio-allantoic membrane treated with heat-inactivated influenza virus (1). This factor transferred antiviral resistance to other virus-susceptible cells. A similar factor had been reported shortly before by Nagano and Kojima (2). The transferred antiviral resistance was not specific for the inducing virus and thus explained previous observations that one virus, be it infectious or inactivated, inhibited the growth of another virus.

K. Loening (Editor)
The Terminology of Biotechnology:
A Multidisciplinary Problem

Even prior to purification, it was clear that more than one type of interferon existed. Interferons were referred to as Type I and Type II, depending on stability at pH 2 and the producing cell. Type I interferons were those produced by leukocytes, lymphoblastoid cell lines, and fibroblasts, primarily in response to virus or double-stranded RNA, and were generally stable to acid. Type II interferon referred to acid labile interferon produced by leukocytes in response to antigen or lectin stimulation. Antigenic differences led to the evolution of the names leukocyte, fibroblast, and immune interferons (3-5 and references therein).

TABLE 1 - INTERFERON NOMENCLATURE

Early Names	Stability	Source	Primary Inducers	Current Names
Type I Leucocyte	Acid stable	leukocytes	virus	Interferon-alpha IFN-α1, α2, α3... IFN-αA, αB, αC...
Type I Fibroblast	Acid stable	fibroblasts	virus dsRNA	Interferon-beta IFN-β
Type II Immune	Acid labile	T-lymphocytes	antigen PHA	Interferon-gamma IFN-γ

With the development of improved induction procedures and new biological and chemical technologies, the interferon field lurched ahead in the late 70's. Development of sensitive protein detection methods and the application of HPLC to protein purification permitted the first purification of interferon to homogeneity (6,7). The availability of small amounts of purified interferon made determination of the amino acid composition and limited sequence information possible (8-10). Because the interferons are among the most potent molecules known, detection of antiviral activity produced in cell-free and oocyte translation systems was possible, albeit laborious. cDNA recombinant clones containing the interferon-coding sequences were identified (11-15) and verified by determination of biological activity, hybridization to interferon specific probes and comparison of the nucleotide sequence with the partial amino acid sequence of purified interferon isolated from induced cells (8-10,16,17). Monoclonal antibody production facilitated the purification and detection of recombinant interferons (18,19). As a result of the rapid application of new technologies to the interferon field, the complete amino acid and nucleotide sequences of human fibroblast and several leukocyte interferons were known in the early 80s (11-15).

1980 and Beyond

After interferon was cloned, it became possible to define the interferons not only by the parameters of stimulus, cell stimulated and antigenicity, but also by sequence and biological activity. An Interferon Nomenclature Committee consisting of twelve leading scientists was organized by the National Institute of Allergy and Infectious Diseases (NIAID). This committee recommended the names interferon-alpha (IFN-α), interferon-beta (IFN-β), and interferon-gamma (IFN-γ) when referring to the pure species that comprise the predominant class of leukocyte, fibroblast and immune interferons, respectively (Table 1). These recommendations were announced in numerous key scientific journals in 1980 and attained worldwide acceptance among the scientific community [see Nature (London) 286:110(1980)].

Both IFN-α and IFN-β possess a range of diverse activities associated with the immune response (reviewed, 20). In addition to potent antiviral activity, these include modulation of cell differentiation, enhanced expression of major histocompatibility class (MHC) I antigens, modulation of antibody and prostaglandin biosynthesis, and modulation of oncogene expression. Species specificities differ. Dose response curves for these various activities also differ. Relative to IFN-α and IFN-β, IFN-γ displays orders of magnitude lower antiviral activity but is a potent modulator of cell differentiation, inducer of Class I and II MHC antigens, and inhibitor of tumor cell proliferation, among other activities (reviewed, 20,21). Although many activities are shared by all interferons, interferons can contain significant differences in their amino acid sequences (reviewed 21,22). Homology between IFN-α and IFN-β is about 35%. IFN-γ is only 7 - 9% homologous with IFN-α and IFN-β. Even by allowing gaps to optimize homology, the extent homology is 15% at best. Further, IFN-α and IFN-β bind to common receptors different from those that bind IFN-γ. It is likely that if IFN-γ and its diverse activities and relatively weaker antiviral activity had been discovered in the 1980s, it might not have been given the name interferon.

By the early 80's a number of groups had screened human genomic libraries of bacteriophage clones in search of DNA segments with homology to IFN-α mRNA (11-15). A striking feature of each α and β interferon clone obtained was the absence of intervening sequences. Several labs found one unique sequence for IFN-β, while numerous sequences of interferon-α's were reported. Discounting what was presumed to be allelic variants, specifically genes differing in 1% or less of their nucleotide positions, over 20 non-allelic genes were identified. These were named IFN-α1, IFN-α2, IFN-α3, etc. or IFN-αA, IFN-αB, IFN-αC, etc., depending on the originating lab. A few of these were shown to be virtually homologous. No single sequence predominated, although it became generally accepted that IFN-α1 (IFN-αD) and IFN-α2 (IFN-αA) represented a major fraction of lymphoblastoid IFN preparations (reviewed, 22).

Among the genomic clones isolated by hybridization with IFN-α DNA were a number of gene-like sequences that could not code for functional interferons due to the presence of one or more stop codons that would result in formation of very short proteins. The significance of these pseudogenes, designated by the prefix Ψ, remains unclear but is generally found among multigene families.

In view of new information available on the sequence of interferon genes, the IFN Nomenclature Committee met in March 1982 at Squaw Valley during a UCLA Symposium on the Chemistry and Biology of Interferons. Recommendations were developed for nomenclature of IFN genes, IFN proteins and IFN preparations. These were distributed for comment through a now-defunct newsletter entitled Interferon Scientific Memoranda and final recommendations were announced in major journals [see J. Immunol. 132:1058 (1984)]. The committee recommended that (i) a prefix, such as Hu for human, be used to designate the animal from which the gene was derived, (ii) a new Greek letter designation would be given to new IFN species if the encoded amino acid residues differed by more than 50% from existing IFNs, and taking into consideration antigenic and biological data, (iii) non-allelic IFN-α genes should be designated IFN-α1, IFN-α2, etc. (iv) natural IFN preparations should be described by the species of origin, producing cell and inducer [such as Hu IFN (Namalwa/Sendai)], and undefined preparations should not be designated by IFN-α, IFN-β, or IFN-γ, and (v) interferon genes should be designated _IFN-α_, _IFN-β_, and _IFN-γ_ (italicized or underlined) [not to be confused with symbols proposed for human gene nomenclature and mapping (23)].

A number of recombinant hybrid IFN-α's have been constructed. These molecules contain additions, deletions, substitutions, and overlaps. They are most appropriately designated by standard nomenclature recommended by the IUPAC/IUB Joint Commission on Biochemical Nomenclature.

INTERLEUKIN-6/INTERFERON-BETA-2/26 kDA PROTEIN/...

The Problem

Raj and Pitha found that induction of human fibroblasts with double-stranded RNA resulted in synthesis of IFN-β and additional proteins ranging from 15,000-70,000 daltons (24,25). In the late 70s, Michel Revel and colleagues cloned a 1.3 Kb messenger RNA species from fibroblasts induced to produce IFN-β. This mRNA translated in a cell-free system into a 26 kDa protein which possessed antiviral activity and was neutralized by polyclonal antisera to an IFN-β preparation (26,27). It was named IFN-β2 by Sehgal and Sagar (28). A group led by Walter Fiers and Jean Content, who were the first to clone IFN-β (29), also observed the 1.3 Kb message which they translated into a 26 kDa protein (30). However, this product was devoid of antiviral activity and was distinct from IFN-β by a number of criteria. The lack of any significant antiviral activity was confirmed by other laboratories using recombinant material (31-33). In 1986, Fiers and colleagues published the first complete sequence of the 26 kDa protein (34).

Also in 1986, Kishimoto and colleagues discovered that B-cell stimulatory factor, or BSF-2, first identified in 1982, was identical in sequence to the 26 kDa protein (35). Experiments by Poupart, Content and co-workers demonstrated that the 26 kDa protein modulated both B-cell differentiation and growth (33). They proposed the name interleukin-6 (IL-6) to eliminate confusion in nomenclature. A subsequent discovery by Gauldie and coworkers in 1987 demonstrated that IL-6 also corresponded to the "hepatocyte-stimulating factor," or HSF, a molecule known since 1983 (36). Finally, cloning and sequencing of the (murine) molecule known as interleukin HP-1, for hybridoma plasmacytoma-1 or plasmacytoma growth factor [PCGF], demonstrated it to be IL-6 also (37).

Thus, by 1988, a variety of activities had been ascribed to the molecule known as IL-6/interferon-beta-2/26 kDa protein/B-cell stimulating factor-2/HSF/PCGF (reviewed, 38). A variety of cells were found to produce this factor, including fibroblasts, monocytes/
macrophages, T-cells and keratinocytes. Inducing agents included LPS, IL-1, TNF, and IL-2. The factor had been reported to act on numerous cells as well, including B-cells, T-cells, macrophages, fibroblasts, hepatocytes, and others.

The Solution

In December 1988, the NIAID convened a nomenclature meeting in conjunction with the New York Academy of Sciences meeting on this factor (38). Reasons for retaining the name IFN-β2 were based primarily on the historical claim of IFN-β2 being its first name and the, albeit extremely low, level of antiviral activity claimed by two groups. However, the absence of strong structural, functional or antigenic relatedness with IFN-β argued effectively against the interferon name. In addition, the gene for this factor was known to contain introns and be located on a different chromosome than IFN-β.

It became clear that a name based on one of its multitude of functions would not be widely accepted and might be misleading. The IL-6 name was considered appropriate since the molecule is produced by leukocytes, although not exclusively, functions during inflammatory or immune response and has a known sequence. Perhaps more importantly, the name IL-6 was found to be the most neutral name. All involved parties agreed to the primary designation of IL-6 with the condition that investigators could use additional synonyms when referring to particular products derived from the interleukin-6 gene (39).

UNRESOLVED NOMENCLATURE ISSUES

Other nomenclature issues in the interferon field remain unresolved at this time. For example, IFN-α related proteins of lengths signficantly greater than the 166 amino acids of human IFN-α are produced by certain species. These interferons may represent a new class of interferons. The names interferon-ω and IFN-αII have been suggested. These and other nomenclature issues will have to be resolved by some venue, which might appropriately be the annual meeting of the International Society for Interferon Research.

A complex array of overlapping activities possessed by numerous cytokines, including the interferons, is now known. The current trend in nomenclature seems to be to give each newly discovered cytokine the IL designation followed by a number. However, in many cases the seminal activity may not be communication between leukocytes. Usually, these factors are multifunctional, are produced by different cell types and have multiple target cells. Furthermore, numerous disciplines are often involved. Non-immunologists seem to be growing wary of extending the IL designation with the discovery of every new cytokine, some of which have major roles outside the immune system. The development of a standardized naming system for cytokines by researchers from a variety of disciplines might assure consistency and facilitate greater ease and accuracy in communication.

REFERENCES

1 Isaacs A, Lindenmann J (1957) Proc R Soc Lond Ser B 147:258
2 Nagano Y, Kojima Y (19858) Cr Soc Biol 152:1627
3 Stewart WE II (1979) The Interferon System. Springer-Verlag, Wien-New York
4 Havell EA, Yip YK, Vilcek J (1977) Arch Virol 55:121
5 Pestka S, Baron S (1981) Meth Enzym 78:3
6 Rubinstein M, Rubinstein S, Familletti PC, Miller RS, Waldman AA, Pestka S (1979) Proc Natl Acad Sci USA 76:640
7 Rubinstein M, Levy WP, Moschera JA, Lai C-Y, Hershberg RD, Bartlett RT, Pestka S (1981) Archs Biochem Biophys 210:307
8 Zoon KC, Smith ME, Bridgen PJ, Anfinsen CB, Hunkapiller MW, Hood LE (1980) Science 207:527
9 Knight E, Jr, Hunkapiller MW, Korant BD, Hardy RWF, Hood LE (1980) Science 297:525
10 Levy WP, Shively J, Rubinstein M, Del Valle U, Pestka S (1980) Proc Natl Acad Sci USA 77:5102
11 Goeddel DV, Yelverton E, Ullrich A, Heyneker HL, Miozzari G, Holmes W, Seeburg PH, Dull T, May L, Stebbing N, Crea R, Maeda S, McCandliss R, Sloma A, Tabor JM, Gross M, Familletti PC, Pestka, S (1980) Nature 287:411
12 Streuli M, Nagata S, Weissmann C (1980) Science 209:1343
13 Taniguchi T, Ohno S, Fujii-Kuriyama Y, Muramatsu M (1980) Gene 10:11
14 Derynck R, Content J, De Clercq E, Volckaert G, Tavernier J, Devos R, Fiers W (1980) Nature (London) 285:542
15 Goeddel DV, Shepard HM, Yelverton E, Leung D, Crea R, Sloma A, Pestka S (1980) Nucleic Acids Res 8:4057
16 Zoon KC (1981b) In: De Maeyer E, Galasso G, Schellekens H (eds) Biology of the Interferon System. Elsevier/North Holland
17 Shively JE, Del Valle U, Blacher R, Hawke D, Levy WP, Rubinstein M, Stein S, McGregor WC, Tarnowski J, Bartlett R, Leé D, Pestka S (1982) Anal Biochem 126:318

18 Staehelin T, Hobbs DS, Kung H-F, Lai C-Y, Pestka S (1981) J Biol Chem 256:9750
19 Staehelin T, Hobbs DS, Kung H-F, Pestka S (1981) In: Colowick S and Pestka S (eds) Methods in Enzymology 78:505. Academic Press, New York
20 Pfeffer LM (ed) (1987) Mechanisms of Interferon Actions Volume II. CRC Press, Boca Raton
21 Grossberg SE (1987) In: Pfeffer LM (ed) Mechanisms of Interferon Actions Volume I. CRC Press, Boca Raton
22 Langer JA, Pestka S (1985) Pharmac Ther 27:371
23 Shows TB, McAlpine PJ (1982) Cytogenet Cell Genet 32:221.
24 Raj NBK, Pitha PM (1980) Proc Natl Acad Sci USA 77:4918
25 Raj NBK, Pitha PM (1980) Nucleic Acids Res 8:3427
26 Weissenbach J, Zeevi M, Landau T, Revel M (1979) Eur J Biochem 98:1
27 Weissenbach J, Chernajovsky Y, Zeevi M, Shulman L, Soreq H, Nir U, Wallach D, Perricaudet M, Tiollais P, Revel M (1980) Proc Natl Acad Sci USA 77:7152
28 Sehgal PB, Sagar AD (1980) Nature (London) 288:95
29 Derynck R, Content J, De Clercq E, Volckaert G, Tavernier J, Devos R, Fiers W (1980) Nature (London) 285:542
30 Content J, De Wit L, Pierard D, Derynck R, De Clercq E, Fiers W (1982) Proc Natl Acad Sci USA 79:2768
31 Reis LF, Le J, Hirano T, Kishimoto T, Vilcek J (1988) J Immunol 140:1566
32 Hirano T, Matsuda T, Hosoi K, Okamo A, Matsui H, Kishimoto T (1988) Immunol Lett 17:41
33 Poupart P, Vandenabeele P, Cayphas S, Van Snick J, Haegeman G, Kruys V, Fiers W, Content J (1987) EMBO J 6:1219
34 Haegeman G, Content J, Volckaert G, Derynck R, Tavernier J, Fiers W (1986) Eur J Biochem 159:625
35 Hirano T, Yasukawa K, Harada H, Taga T, Watanabe Y, Matsuda T, Kashiwamura S, Nakajima K, Koyama K, Iwamatsu A, Tsnunasawa S, Sakiyama F, Matsui H, Takahara Y, Taniguchi T, Kishimoto T (1986) Nature (London) 324:73
36 Gauldie J, Richards C, Harnish D, Lansdorp P, Baumann H (1987) Proc Natl Acad Sci USA 84:7251
37 Van Snick J, Cayphas S, Szikora J-P, Renauld J-C, Roost EV, Simpson RJ (1988) Eur J Immunol 18:193
38 Sehgal PB, Grieninger G, Tosato G (eds) (1989) Annals NY Acad Sci Vol 557, New York Academy of Sciences, New York
39 Paul WE, Laughlin CA, Johnston MI (1989) Annals NY Acad Sci 557:579

The Nomenclature of Monoclonal Antibodies

L. Blaine

American Type Culture Collection
12301 Parklawn Drive, Rockville, MD 20852-1776, U.S.A.

INTRODUCTION

Hybridoma technology is a relatively new tool for the development of biochemical reagents having wide applicability in many areas of biology and chemistry. These reagents are termed "immunoreagents" because they are produced by cultured cell lines selected for their ability to secrete immunoreactive substances. The most common of these immunoreagents are monoclonal antibodies (MABs), which exhibit complex reactivity patterns although they are specific to unique epitopes. Epitopes are antigenic determinants of specific structure, i.e., chemically active structural components of proteins and complex carbohydrates which interact with antibodies (immunoglobulins). MABs developed through hybridoma technology are highly specific to individual epitope topologies, but because similar epitopes can be found on diverse proteins and polysaccharides, the MABs may react with dissimilar cell types or biochemical substances, despite their definition as "reagents of defined specificity".

The biological terminology required to describe the characteristics of these immunoreagents spans virtually all of the subdisciplines of biology and organic chemistry. MABs can be generated which react with specific epitopes on molecules which comprise the organic structure of all types of organisms of the plant and animal kingdom. Terminology used to describe MABs covers the taxonomic nomenclature used for each genus, species, and strain of organism with which MABs react, as well as the cells, subcellular structures, organs, tissues, and biochemical substances which make up these organisms.

Monoclonal Antibodies (MABs) and the cloned cell lines from which they arise are not named, as entities, according to defined rules.

K. Loening (Editor)
The Terminology of Biotechnology:
A Multidisciplinary Problem

In fact, a very confusing situation exists which often results in misidentification of these immunoreagents. Identical products handled by various distributors, because they do not carry unique identifiers, may be assigned a wide variety of product names. On the other hand, disparate products generated in various laboratories throughout the world are often given identical designations.

As the number of MABs continues to grow exponentially, the situation will become even more confusing. The wide applicability of MABs in many fields of science and the availability of newer and less costly methods of generating these highly specific immunoreagents insures the continued increase in their number and sources. Clearly, some system is required to alleviate the current confusion. Several proposals have been offered to standardize the nomenclature of MABs. This paper will point out some of the impracticalities of the proposed solutions and offer alternatives.

I. Nomenclature Schemes For Monoclonal Antibodies

The most common suggestion for naming of MABs is to give each laboratory generating such reagents an acronym. This acronym would be affixed to the name of all MABs produced by the laboratory along with some other unique identifier which distinguishes each MAB generated by that laboratory. The system of assigning acronyms, of course, is not original. Airports of the world have long been identified by acronyms and in the field of biological sciences the World Data Center for Microorganisms assigns acronyms for all participating culture collections. Both of these examples represent a finite number of organizations which are linked by common goals and a structured oversight body to assign and maintain lists of acronyms. These seem to be basic elements for the success of such a system. In the case of innumerable and diffuse laboratories generating MABs, however, there is no linkage among them and certainly no oversight body to handle the use of acronyms. Furthermore, the generation of MABs is usually an incidental part of the work of each laboratory. MABs are generated as tools for some experimental purpose and there is little interest in systematized naming conventions. While the laboratory may publish on a particular MAB or share it with other investigators, these reagents are usually not given to repositories for cataloging or control. This solution, therefore, seems impractical at this time.

Table 1 illustrates a classification scheme proposed by A. Bussard (1) for the various derivations of MABs. If the Class Number were made an integral part of the name of the MAB it would disclose some information about the method of developing the product and also some genomic information on the hybridoma or other immunoclone producing the MAB. The primary purpose of this classification scheme is to differentiate the various immunoreactive products of cloned cell lines and is, therefore, not restricted to MABs. If adopted, however, it would help to distinguish unique MABs based on the methodology of their generation. It would not totally eliminate duplication of names unless used in combination with other standardized naming conventions.

Table 1

CLASSIFICATION OF IMMUNOCLONES

#	Type	Genome	Immortalization procedure	Transfer of immunological information	Product
1	BP	2	Hybridization	-	MAb
2	B_1B_2P	3	Double hybridization	-	MAb Hy
3	B_1PB_2P	4	Triple hybridization	-	MAb Hy
4	B	1+	Virus transfection	-	MAb
5	P	1+	None	DNA transfer	MAb
6	P	1++	None	Multiple DNA transfer	MAb Hy

A third proposal has been offered by H. Baumgarten and K. Kurzinger (2) who describe a nomenclature scheme used at Boehringer Mannheim for several years. This proposal suggests that the best way of defining MABs unambiguously is to embed information about their reactivities and methods of preparation into the name. Several examples of this naming convention follow.

Figure 1 illustrates the general principles of the proposed scheme, i.e., the name would contain at least the type of antibody, the immunogen used to stimulate production of the antibody, the animal used to raise the antibody, and the isotype of the antibody.

Figure 1

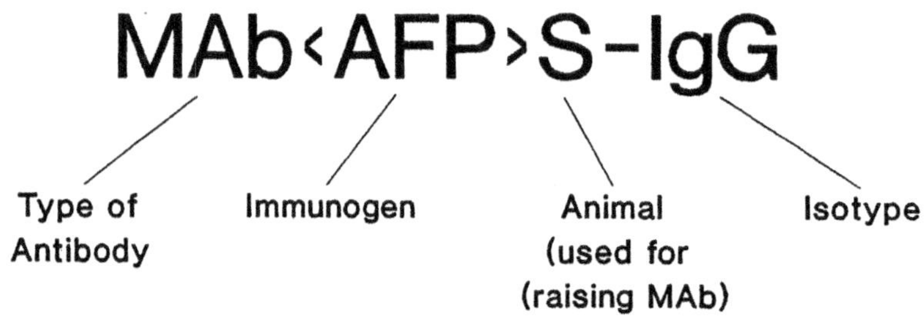

The examples given in Figure 1 contain fairly recognizable components, i.e., MAB=monoclonal antibody, AFP=alpha fetoprotein, S=sheep, and IgG=immunoglobulin G. In the following examples, the components may not be quite as clear.

The character string "BAb<H-IgA,POD>M" represents the name of a bispecific mouse monoclonal antibody, reacting with both human Immunoglobulin A and Peroxidase. "MAb<H-CEA>M-Tu2-Fab(DE-<M-Fcg>)-Gal" refers to a conjugate of Beta-galactosidase and the Fab fragment of the monoclonal antibody Tu2 which is directed against human carcinoembryonic antigen. Fab was separated from Fc gamma by DEAE chromatography.

The authors provide lists of abbreviations which they have adopted in their naming scheme. The use of "M" for mouse or "H" for human is fairly straightforward, but their suggestion to use "Ca" for canine and "F" for feline could cause confusion because of inconsistency in format and inadvertent mixing of nouns and adjectives as descriptors. The addition of non standard abbreviations for purification procedures, e.g., "CM" for carboxymethyl cation exchange chromatography or "IS" for positive immunosorption could result in totally unintelligible names for MABs, especially to scientists whose primary language is not English. This scheme, which may be entirely suitable for naming certainly break down if applied to the hundreds of thousands of MABs which now exist.

II. The Terminology Of Biology

MABs react with macromolecular substances present in virtually every organism of the biosphere. Any haphazard attempt to list and/or abbreviate the biological entities with which MABs react and incorporate them into the names of these immunoreagents is bound to fail. Members of the Biological Unions of the International Council of Scientific Unions (ICSU) spend entire careers on the naming and classification of organisms, organs, cells, tissues, diseases, and macromolecular substances which contain the binding sites for MABs. The Committee on Data for Science and Technology (CODATA) of ICSU has recently created a Commission on the Terminology & Nomenclature of Biology whose mission is to facilitate and publicize the work of the terminological and nomenclature committees of the biological unions. The publication of a comprehensive list of biological terminology which would be relevant to the naming of MABs is not a trivial undertaking.

A. The Hybridoma Data Bank Content and Structure: In 1983, CODATA and the International Union of Immunological Societies (IUIS) established an international Hybridoma Data Bank(HDB) to provide a comprehensive directory of information on hybridomas and other cloned cell lines and their immunoreactive products, such as MABs. Data for this resource are collected by five "Nodes" or data collection centers located in various parts of the world. All of the Nodes used standardized format and terminology to store and exchange data. Data Bank records contributed by each collecting center are subjected to stringent validation procedures to insure that the biological terminology is consistent and follows guidelines set by nomenclature standardization committees whenever possible.

The need for guidelines in choosing standard descriptors for biological entities is becoming more critical because of the recent developments in biotechnology and the resulting increases in interdisciplinary research. The HDB provides an excellent means for providing standardized comprehensive descriptions of MABs. Since each MAB listed in the HDB carries a unique identifier composed of a random numeric sequence, this numeral (possibly coupled with a laboratory designated name) could serve as an appropriate naming scheme for MABs and other products of immunoclones. After generation and characterization of a new MAB, investigators could submit the data to the Bank and be given a unique identifier. The complete data record containing known information about the MAB could then be made available to all. The HDB unique identifier could be used in the literature and in other references to a specific MAB. Thus a reference to MAB 00234, reactive to human alpha fetoprotein or MAB 00235 reactive to several tumor antigens would serve as a simple pointer to the complete description as listed in the HDB.

A simplified overview of the structure and content of the HDB is presented below to point out that the characteristics of MABs represent a paradigm for the problems of biological terminology in general. It seems reasonable to consider the option of naming products of biotechnology, such as MABs, genetically engineered organisms and proteins, cloned cell lines, etc., not by elaborate naming schemes which seek to convey descriptive characteristics of these products, but by assigning unique identifiers which link them to data bank records containing complete and standardized descriptions.

Figure 2

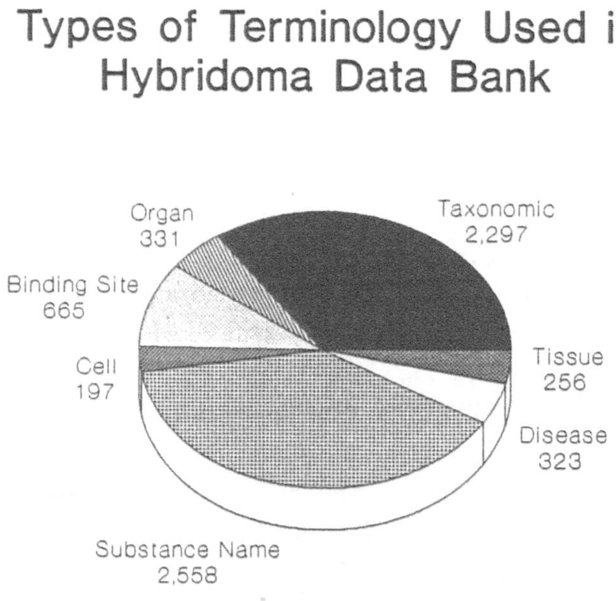

Figure 2 illustrates the relative number of terms which now exist in the HDB from some of the major categories of descriptors mentioned above. It can be observed that from a list of over 10,000 controlled descriptors, the highest number are biochemical substances, i.e., macromolecules and specific fragments of these macromolecules which define the epitope specificity of the MABs.

Table 2 illustrates the sections of the controlled vocabulary used to code a database record on a MAB which reacts with swine neurofilament protein. It can be seen that terminology from the taxonomic codes for organism names (Sus scofa), and controlled terminology for organ names (spinal cord), subcellular organelles (neurofilament), and substance names (neurofilament protein) were used to code this record. The hierarchical structure of the controlled vocabulary places similar types of biological terms in close proximity to one another and allows the database searcher to group similar terms for data retrieval. Note numeric codes which are used to group synonyms and allow for "range searching" of related terms.

Table 2

SAMPLE HDB RECORD

```
UNIQUE IDENTIFIER: 10483
DESIGNATION: C>NR4 ;developer
IMMUNOGEN:  G>Sus scrofa CN>swine O>spinal cord
     U>neurofilament  SN>neurofilament protein
DONOR:  G>Mus musculus CN>mouse
PROCEDURE: ;fusion
PRODUCT: ;IgG1
REACTANT: CE>neuron U>neurofilament
     SN>neurofilament protein  MW>68 kD
     G>Sus scrofa CN>swine
```

CONTROLLED VOCABULARY OF BIOLOGICAL TERMS

0050700-0050799 NERVOUS SYSTEM

```
0050700: nervous system (unspecified)
0050701: brain (syn. cerebrum)
0050702: cerebellum
0050703: medulla oblongata (syn. medulla)
0050704: central nervous system
0050705: hypothalamus
0050706: thalamus
0050707: corpus callosum
0050709: nerve cord
0050710: spinal cord
0050711: sympathetic ganglion
```

0050712: nodose ganglion
0050715: dorsal root ganglion (syn. spinal ganglia)
0050716: nodes of Ranvier
0050717: brain temporal cortex (syn. cerebral temporal cortex)
0050718: brain frontal cortex (syn. cerebral frontal cortex)
0050719: brain cortex (syn. cerebral cortex)
0050720: peripheral nerve
0050721: cerebellum cortex
0050723: cerebral lobe
0050725: optic lobe
0050726: optic tectum
0050727: ciliary ganglion
0050728: ganglion

0081650-0081750 CYTOPLASM

0081650: cytoplasm
0081653: cytosol
0081655: cytoskeleton (syn. cytoplasmic filament)
0081656: non-granular cytoplasm
0081657: peribacteroid membrane
0081660: intermediate filament (see also neurofilament, tonofilament, microfilament)
0081661: neurofilament
0081662: tonofilament
0081663: microfilament
0081664: stress fiber
0081665: microspike
0081666: glial filament
0081667: P granule
0081669: cortical secretory vesicle

0500241-0500265 CYTOSKELETAL PROTEINS

0500241: intermediate filament protein
0500250: glial fibrillary acidic protein
0500250 ($ GFAP coded as glial fibrillary acidic protein)
0500252: intermediate filament associated protein
0500254: neurofilament protein
0500255: desmin
0500257: vimentin

TAXONOMIC CODES FOR ORGANISM NAMES

Family	Suidae	8574000-8574490
8574000:	Suidae	(pigs)
8574020:	Sus sp.	pig, swine
8574030:	Sus scrofa	pig, swine

The primary purpose of the HDB is to store and manage comprehensive information on hybridomas and MABs, including methods and immunizing agents used for development of the hybridomas, immunoreactive products, reactivities, and known non-reactivities of the MABs, and the availability of the cloned cell lines and their products. Thus, the HDB provides a directory service for locating immunoreagents with specific characteristics.

The data bank can also be used as a research tool for structural and taxonomic analyses. Structural similarity among macromolecules and taxonomic relationships among organisms, based on their reactivity patterns with MABs has been demonstrated by analysis of data in the HDB.

Computer programs can be used to perform cluster analysis of the structural similarities among macromolecules based on their reactivity patterns with MABs. Molecules which have similar epitopes will "cluster" when subjected to taxonomic analysis based on their reactivity patterns with a defined set of MABs. The output of cluster analysis is a list of which molecules group closely together based on similar properties. The nomenclature and terminology contained in the HDB is, therefore, truly a reflection of the nomenclature of biotechnology.

The use of databases such as the HDB as tools for scientific research demands a precise and comprehensive tool which encompasses all of the terminological work produced by the Biological Unions and other sanctioned nomenclature groups. Historically, the various components of ICSU, specifically the Biological Unions such as the International Union of Biological Sciences (IUBS), the International Union of Biochemistry (IUB), International Union of Immunological Societies (IUIS), International Union of Microbiological Societies (IUMS), and others have appointed committees to recommend standardized nomenclature and terminology in defined subdisciplines. The IUB, for example, has a committee which sets standards for enzyme nomenclature. Their publication(3) is widely recognized and used by authors publishing in the biological literature, by database developers building thesauri of biological terms, and by searchers of biological databases. Unfortunately, the situation regarding enzyme nomenclature is somewhat unique. Although other groups have managed to develop internationally recognized nomenclature systems within specific areas of biology, the existence of standards is the exception rather than the rule.

There are numerous organizations involved in classifying objects and developing nomenclature for the classes. All the major branches of biology (zoology, botany, bacteriology, virology) have nomenclatural committees which deal with the naming of taxa at the species level and above. Two major problems exist in utilization of the products of these committees; a) disagreement among various factions which make up the committees, resulting in confusion for potential users of the nomenclature, and b) lack of reliable pathways to the products of these committees.

Developing standards for the naming of organisms, the traditional role of taxonomists, is only the beginning of the problem. A far more complex aspect of biological nomenclature is the development of well defined and standardized terminology for describing the morphological, physical, and biochemical components and characteristics of these organisms.

Building a thesaurus of controlled terminology for the HDB has been extremely labor intensive because of the lack of standard terminology in many areas of biotechnology. In a factual database such as the HDB, the terminology must be very precise. To prepare a database record, for example, describing the reactivity of a MAB to a particular protein such as gp70 (a specific viral protein designated by its molecular weight), one cannot "index" this term under the broader term "glycoproteins", as might be done in a bibliographic database. It must be entered into the database in its precise form. Table 3 provides a partial listing of other glycoproteins which are named simply by their molecular weight. Names such as these are quite useless without other information which further characterizes the proteins.

Table 3

Examples of Glycoproteins Named by Molecular Weight

0001582: gp41	0001603: gp70b
0001583: gp25	0001604: gp70c
0001584: gp51	0001607: gp70-p15E
0001585: gp52	0001615: gp80
0001586: gp42	0001620: gp85
0001587: gp54/70	0001627: gp90
0001588: gp68	0001630: gp110
0001600: gp69/71	0001635: gp160
0001601: gp70	0001636: gp180/90
0001602: gp70a	

The problem of protein nomenclature is especially complex, not only because proteins are often named by their molecular weight, as in the case of gp70, but that they may also be named based on their reactivities with antibodies, location within a cell or on the surface of a cell, or by function. Identical proteins may, in fact, be given different names simply because they were identified by different types of research methods. Integration and comparison of the data stored in various types of factual databanks such as sequences, structures, and antibody binding sites ar necessary to obtain a complete picture of similarities and differences among proteins. Access to a universally accepted nomenclatural tool which can be constantly updated regarding synonymy and history of protein names would be a great help to database producers. It would also help to link the information in the various biotechnology databases together and allow researchers to obtain a complete and integrated data set on any topic of interest.

Because of the diversity of biochemical substances with which MABs react, HDB coders must cope with biological terms which are rather loosely defined as illustrated in Table 4. Classes of these substances include cell surface antigens, receptors, tumor antigens, and developmental antigens, Fortunately, because of the work of organized nomenclature groups working within the various subdisciplines of biology, database producers also have a group of standardized, well defined terms with which to work. Classes of these substances include blood coagulation factors, hormones, enzymes, histocompatibility antigens (mouse and human), leukocyte surface antigens, and hepatitis antigens. Examples of this type of biological terminology are illustrated in Table 5.

Table 4

Examples of Poorly Defined Biochemical Substances

RECEPTORS

elastin receptor
elastin receptor complex
extracellular matrix receptor I
extracellular matrix receptor II
beta adrenergic receptor
beta 1 adrenergic receptor
beta 2 adrenergic receptor
muscarinic receptor
nicotinic acetylcholine receptor
bungarotoxin-acetylcholine receptor complex
dopamine D2 receptor
gamma-aminobutyric acid/benzodiazepine receptor complex
adenosine receptor A1
adenosine receptor A2
opioid receptor
opioid delta receptor

opioid kappa receptor
opioid mu receptor

TUMOR ANTIGENS

leukemia associated antigen p24
adenocarcinoma related antigen
osteosarcoma related antigen
TSP-180
pancreas carcinoma associated antigen
prostate carcinoma associated antigen
T antigen
small T antigen
middle T antigen
large T antigen
large T-related antigen
non-viral T antigen

Table 5

Well Defined Biochemical Substances

BLOOD COAGULATION FACTORS (4)

Factor V	Prekallikrein
Factor VII	Platelet Activating Factor
Factor VIII	Platelet Factor 3
Factor IX	Platelet Factor 4
Factor X	Platelet Membrane
Factor XI	Glycoprotein
Factor XII	Prothrombin
Factor XIII	Thrombin
Fibrinogen	Thromboplastin
Kallikrein	Von Willebrand Factor

HISTOCOMPATIBILITY ANTIGENS (5)

H-2	H-2s
H-2Db	H-2b
H-2Dk	H-2u
H-2Kb	H-2d
H-2Kr	H-2Kq
H-2Kd	H-2a
H-2Dd	H-2Dq
H-2Kk	H-2v
H-2q	H-2t1
H-2k	H-2t2
H-2p	H-2b/d
H-2r	H-2Ld
H-2f	H-2Lq

LEUKOCYTE SURFACE ANTIGENS (6)

(Cluster Designations)

CD1	CD3a
CD1a	CD3b

CD1b
CD1c
CD2
CD2a
CD2b
CD2c
CD3
CD3c
CD4
CD5
CD6
CD7
CD8

The need for standardized terminology within the biological sciences has been recognized for many years. This need is growing more critical because of the recent proliferation of automated systems for storage and retrieval of biological information.

The nomenclature and terminology problems discussed in this paper are not unique to the MABs. All biological database developers are facing problems in choosing standard descriptors for biological entities. The problem is more pronounced due to the recent increase in interdisciplinary research. Cross disciplinary retrieval of data becomes increasingly difficult as the "vocabulary of biology" proliferates. Traditionally, nomenclature committees have existed within various biological fields. Their function has been to provide advice and set standards for terminology and nomenclature within subspecialties of the field. Several organizations are now addressing the problem from an interdisciplinary point of view.

As mentioned above, ICSU and CODATA are major contributors to the establishment and dissemination of standards for biological terminology and nomenclature. CODATA's Commission on the Nomenclature and Terminology of Biology addresses the issues on a global and interdisciplinary scale. Others, such as the Matrix of Biological Knowledge Group, are working on development of a "biological matrix" which will aid in drawing generalizations from the vast biological knowledge base. This will require a common data dictionary of biological terms which are standardized and defined.

The U.S. National Library of Medicine (NLM) has an online directory of biotechnology information resources (DBIR) which includes biological nomenclature committees and their publications. NLM's National Center for Biotechnology Information (NCBI) is involved in the development of an online hierarchical dictionary of standardized biological terms. This list represents one activity in NCBI's mission to coordinate efforts to collect and disseminate biotechnology information on a worldwide bases. The International Information Centre for Terminology (INFOTERM) works in liaison with the International Organization for Standardization (ISO) to coordinate terminology standardization. These and other activities should increase accessibility of standardized terminology across disciplines and provide external guidelines which will greatly reduce the work and the inconsistencies currently found in the growing number of factual databases covering the various aspects of biotechnology.

SUMMARY

Monoclonal antibodies (MABs) are named by their developers in an arbitrary manner without benefit of standard nomenclature conventions. The possibility of developing standards has been discussed, but is becoming increasingly remote because of the large number of MABs in circulation and the diversity of laboratories generating these reagents. Widespread applications of MABs have expanded this technology to a multidisciplinary community, thus adding to the difficulty of nomenclature control. A solution to the problem of MAB nomenclature has been proposed and is being implemented by the CODATA/IUIS International Hybridoma Data Bank (HDB). This data bank contains complete records on MABs developed throughout the world. Since each record covers a single MAB with an assigned unique identifier, specific MABs and their characteristics can be linked to this number. If the HDB convention is accepted by authors and journal editors, duplication and non-standardization of the names of MABs will no longer be a significant problem.

<u>References:</u>

1 Bussard A, (1989) Immunoclones 1(1):4-5
2 Baumgarten H, Kurzinger K, (1989) J. Immunol. Methods 122:1-5
3 Webb EC, (ed) (1984) Enzyme Nomenclature: Nomenclature Committee of the International Union of Biochemistry, Academic Press, Inc., New York
4 U.S. National Library of Medicine, (1989) Medical Subject Headings, U.S. Department of Commerce, National Technical Information Service, Washington, D.C., PB-89-10028
5 Steinmetz M., Hood L., (1983) Science 222:727-33
6 Nomenclature Committee of the 4th International Workshop on Human Leukocyte Differentiation Antigens, (1989) Immunology Today 10(8):253-8

Nomenclature Issues in the Human Genome Project

M. D. Zorn and C. R. Cantor

Human Genome Center
Lawrence Berkeley Laboratory
Berkeley, CA 94720, U.S.A.

INTRODUCTION

The recent progress in biotechnology, in particular in techniques to manipulate the genetic molecule DNA, has led to probably the most exciting, and also much debated enterprise in biology: to reveal the sequence of the human genome.

The Human Genome Project: Founded about two years ago, the Human Genome Project consists of a program to map and sequence the human genome, i.e. to construct a contiguous map of the genes and adjoining pieces, and establish the DNA sequence that constitutes the human genome. Within the next five years, the starting period, it is expected that about 50% of the genome will be mapped and 1% will be sequenced. By the end of that term, an improvement of the technology by 5 – 10 fold is conceivable. The remaining ten years of the project will see a gradual change of focus from mapping to sequencing plus continued developments in technology. Beside the time schedule the project falls into three related and most likely parallel themes:

Research in methods and procedures to improve current mapping and sequencing technologies as well as finding new ways to determine the sequence of a DNA molecule, e.g. using scanning tunneling electron microscopy to directly read the sequence off the molecule. Better and faster tools for automating standard laboratory procedures are crucial to complete the project in the announced time frame of about 15 years.

Production efforts to do the actual work of mapping and sequencing. Despite robotic automation and technology improvements there remains still an enormous amount of work to gather all the data. Organizational structures, such as the Human Genome Centers of the U.S. Department of Energy or the National Center for Human Genome Research of the National Institutes of Health, have been set up to guarantee coordinated activities and to ensure that gaps in the maps or sequence are filled.

Analysis of the data, another major research effort, in order to make use of all the information gathered. Improved computational tools to construct and verify maps, to assemble sequences, and to analyze the sequences need to be developed. New approaches to databases and data management are needed to ensure that the information is accessible to the biologists.

K. Loening (Editor)
The Terminology of Biotechnology:
A Multidisciplinary Problem

In contrast to projects of similar size in other scientific disciplines, where scientists gather around few large research centers, the Human Genome Project will be accomplished by a large number of cooperating research facilities. A small number of central research laboratories will ensure the progress of sequencing the entire genome and enable communication among the participating researchers by providing resources for databases and analysis software.

Scope: Besides challenges in the development of new techniques and tools to speed up the data acquisition, issues in data management and nomenclature will become more important. However, the nomenclature problems differ from the use of ordinary biotechnology not so much in their nature, but in their size. The human genome consists of 24 chromosomes (i.e. 22 autosomes and the X and Y sex chromosomes) which contain about 3 billion of DNA monomers, i.e. the four bases Deoxyadenylate, Deoxycytidylate, Deoxyguanylate and Deoxythymidylate. The latest release of GenBank [1], currently the largest repository for DNA sequences, encompasses roughly 30 million bases. So we will need about 100 GenBank's to store the information of only one human genome. However, since more than one different human genome is worked on, we expect to find discrepancies and biological polymorphisms among the data generated by different research groups, so that the real amount of data will be much larger than 3 billion bases.

NAMING PROBLEMS IN MOLECULAR BIOLOGY

Types of Maps

Currently there exist only very low resolution **genetic** maps of the human chromosomes derived from observing the inheritance of certain known traits. A genetic map shows a possible order of genes along the chromosome, but does not reveal any physical distances between them. To improve the resolution and assign physical map units, several methods to obtain **physical** maps have been developed. The DNA molecule is cut into pieces, e.g. by using restriction enzymes, i.e. enzymes that recognize particular sequence patterns and break a phosphodiester linkage, or by physically damaging the DNA with X-rays or sonication. The resulting fragments have to be put into an order. The resolution of the map depends on the size of the fragments. The larger the DNA fragments, however, the easier it is to construct the map. Reasonably small fragments that overlap each other and permit sequencing the DNA are commonly referred to as **ordered libraries**. The ultimate goal is to establish a contiguous DNA **sequence** of the human genome.

Common Landmarks

All kinds of maps may be integrated by using elements that are common among the different types of maps. In biological terms, one can establish a relationship between two different DNA fragments by hybridization. The DNA is isolated from cells, digested into smaller fragments, and then separated according to the size of the fragments using gel electrophoresis. The gel is then transferred to a membrane to fix the banding pattern. The test fragment, a DNA probe — usually a smaller piece of DNA (hundreds of bases) — is isolated in a similar manner and labelled using Phosphorus–32 or chemiluminescent dyes. The membrane containing the separated fragments is then incubated with the labelled DNA. After several washing steps to enhance the significance of the binding of the labelled probe, the membrane is exposed to X-ray film. Bands that appear on the autoradiogram indicate common DNA sequences between the fragments and the probe. (See figure 1) Using this hybridization method, DNA probes can be used as common landmarks to integrate various types of maps.

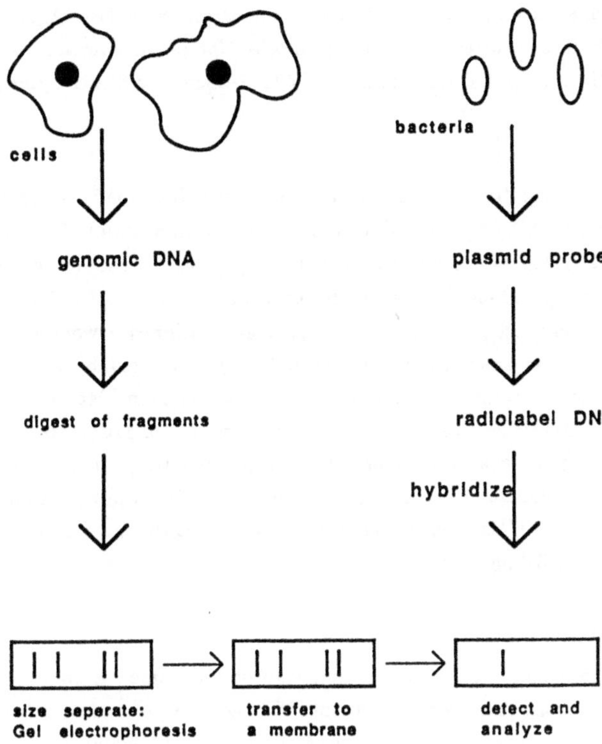

Figure 1: Schematic Illustration of Hybridization

In *in-situ* Hybridization, the probe is labelled with fluorescent markers and hybridized to a metaphase chromosome. The chromosomal location of the probe is investigated under the light microscope, and thereby linked to a specific region or band of the chromosome. Probes also serve as primers for DNA sequencing and to identify human DNA in somatic cell hybrids.

The naming conventions are as different as the methods used. In fact, there exists no common scheme of assigning probe names. Moreover, due to differences in the DNA, polymorphisms, that arise from using different cell lines or studying a different population, the correlation between fragments and identifying DNA probes is even more ambiguous. To keep track of all the different probes and different names for the same probes is an increasingly difficult task. The complex structure definition of the Human Gene Mapping Library (HGML) database [2], which stores information about some kinds of probes, will serve as an example.

Gene Names

Gene names designate DNA fragments which are known to contain genes, i.e. useful information for the cell machinery. Current estimates assume that only about 10% of the coding capacity of the human genome is used to encode genes. Gene names are usually derived from the observed genetic trait (BCEI — breast cancer, estrogen-inducible sequence expressed in), a known disease (AD1— Alzheimer's Disease), or a particular enzyme (SOD1 — superoxide dismutase 1, soluble). To choose an appropriate name for a gene is quite easy when the gene's function is known, but, since there is no general convention, the assigned names are sometimes strange and meaningless for colleagues. Special naming problems arise in cases where a gene occurs in multiple copies on the chromosome (ACTA, ACTB, ACTBP1–6 — actin alpha, beta, and beta pseudogenes), or the gene codes for an enzyme that participates in a common pathway or is itself made of multiple subunits (CYP1 — members of the Cytochrome P450 subfamily).

D-Numbers

D-numbers are used to identify larger DNA fragments whose function is yet unknown, e.g. D21S13. The letter D simply means DNA fragment, which is a redundant information. 21 designates the chromosome. S means single copy DNA as opposed to repetitive sequences; 13 shows the chronology of assignment: it is the 13th single copy probe assigned to chromosome 21. The DNA clone which carries that piece of DNA is usually picked at random from a library and assigned a working name by the laboratory. Through collaborators the clone may get broader publicity under its lab name. When the clone has been mapped regionally to a chromosome, however, the Human Genome Mapping Workshop Nomenclature Commitee assigns a D-number to this clone. Continued characterization may reveal the virtual identity of two separately mapped clones or determine a proper function. In this case the D-number has to be changed. For example, DMD — the Duchenne muscular dystrophy gene includes DXS142, DXS164, DXS206, DXS230, DXS239, DXS268, DXS269, DXS270, and DXS272. This nomenclature does not help in unambiguously establishing a name.

Libraries – Arrayed Libraries

One way to achieve a high resolution physical map of the human genome employs the construction of ordered clone libraries. The DNA is cut into manageable pieces and put into bacterial viruses to propagate them further. The presence of a piece of human DNA is indicated by the formation of virus plaques on a lawn of bacteria. Assuming fragments of 20kb (20,000 bases), about 690,000 virus plaques are necessary to cover the human genome with a completeness of 99%. One chromosome on the average would be contained in about 31,000 plaques. To obtain a map with this method means to establish a consecutive order for all the plaques or 20kb fragments. Each fragment has to be hybridized to all the other fragments to arrange them on the map. This poses not only a considerable amount of practical work, but also a challenge for the data management and naming.

Arrayed libraries [3] have been developed to deal with these high numbers. The clones containing the fragmented DNA are analyzed in arrays of 96, 8 rows and 12 columns. The clones in one row can be pooled and hybridized against the pooled columns of another array to detect similarities. Individual clones of interest are ultimately identified by matching the row with the column which yielded the hybridization signal. Thus the work is reduced in theory from a quadratic to a linear number of hybridizations. In practice, however, two arrays are not simply related in a one to one fashion. A clone in one array might well hybridize to more than one other clone in another array.

SEQUENCE BASED IDENTIFIERS

To overcome the deficiencies described above and other problems the STS — Sequence Tagged Site — approach has been proposed [4].

If there were a homogeneous distribution of the four bases throughout the genome, a consecutive stretch of 16 bases should occur only once in the whole human genome. Because the base distribution is not nearly homogeneous, it is regarded that the chance of finding two identical 21 bases sequences is less than 0.1% except for trivial repeating sequences like 21 Adenylates in a row. Instead of finding different names in multiple naming systems, the STS proposal suggests to use the inherent sequence itself as an unique identifier of every piece of DNA. Every clone, probe, piece of DNA has such an identifier in form of the DNA sequence. With the current techniques this identifier is easy to obtain, sequencing a stretch of a few hundred bases is routine work today. But beyond solving the nomenclature problem this approach makes it unnecessary to maintain huge stock collections, clone libraries, etc. of all the different DNA fragments by taking advantage of PCR — Polymerase Chain Reaction [5].

Polymerase Chain Reaction: PCR uses a complementary set of primers, i.e. they hybridize to opposite strands of DNA and usually face each other, to amplify the DNA that lies between the two primer binding sites. In the first cycle a polymerase, a DNA replicating enzyme, starts at a primer and extends the newly made strand well beyond the binding site of the complementary primer. After denaturation and annealing of a new set of primers, the polymerase generates the complementary strand, but when the template strand is the product of the first cycle, the polymerase stops after the binding site of the first primer, because the DNA ends there. In a series of polymerase/denaturation/annealing cycles the polymerase shuttles between the two primers thus amplifying the DNA right between them.

STS and PCR: The ability to amplify DNA between a set of primers allows one to retrieve every fragment from a human chromosome just by using the appropriate set of primers. The published unique identifier for the DNA fragment of a few hundred bases provides enough information to construct the set of primers. With these primers, clones containing the STS fragment are simply picked by providing a template for amplification and thus generating the STS fragment. There is no need for maintaining biological stocks, no need for days of waiting for the sample requested from a collaborating researcher. The sequence data provide a robust identification and allow the generation of the DNA fragments. Furthermore, as all the other maps have underlying DNA sequences, the maps generated by different mapping methods can be integrated through the use of this common language.

Mapping with Inverse PCR: Primers facing away from each other can be used to amplify a circular DNA molecule [6, 7]. Chromosomal DNA is cut with restriction enzymes. The pieces are circularized using a small linker carrying the binding sites for two outside bound primers. Using PCR the fragment can be amplified and is directly available for sequencing.

C. L. Smith has proposed to use inverse PCR to bridge the gap between the sites of rare cutting restriction enzymes. Sequences surrounding rare cutting restriction enzyme sites are available by analyzing *linking clones* and these could be used to construct appropriate STS's. These rare cutting sites also tend to occur in the vicinity of genes. Thus the idea is appealing in three ways: (1) the initial fragments are small enough to permit inverse PCR and sequencing of the amplified DNA, (2) the STS's themselves will contain the information needed to construct a physical map, and (3) the DNA analyzed is interesting because of the high possibility of nearby genes.

HUMAN GENOME WORKSTATION

The real problem in the Human Genome Project is going to be information overload. Figure 2 shows a comparison of genomes from different organisms in terms of "*phone books*". The human genome would thus comprise 200 books with 1000 pages each containing 500 genes. The largest genome that has been fully sequenced to date is the Human Cytomegalovirus, which is about 220,000 bases or 7 pages long. The huge amount of information expected from the Human Genome Project will be overwhelming without extensive computer support.

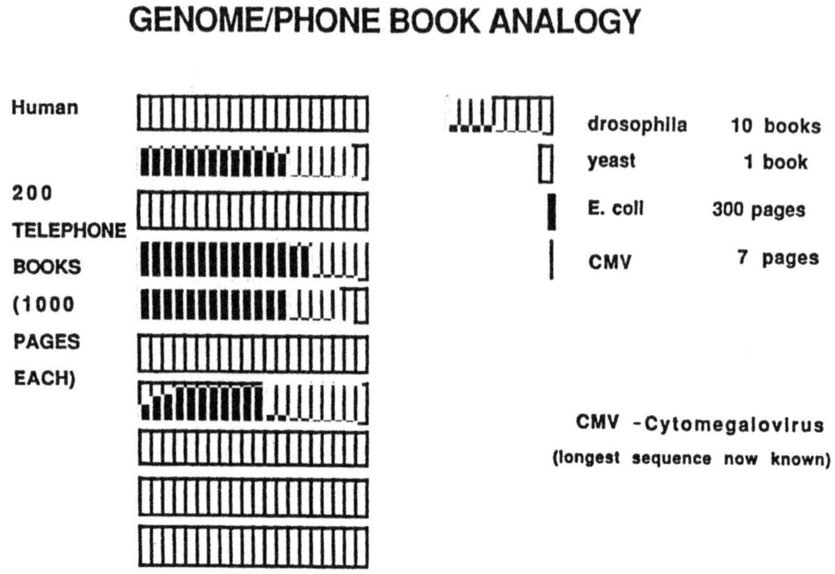

Figure 2: Amount of DNA Sequence Information in Terms of Phone Books

The computer will become a valuable tool for biologists as it is now for other science disciplines. The data will obviously be entered into various databases. A Human Genome Workstation needs to be developed that will address any relevant data base in a manner transparent to the user. Currently each and every database has its own peculiarities, but in order to keep the information accessible, a common interface has to be developed. With the advent of standard windowing systems it should be possible, in the near future, for a user to get the same screen display, the same Look-and-Feel, regardless of the terminal he is using, or the data base he is reading, or the application program he is running. The user may customize his workstation to accommodate local preferences for probe names, cell lines, etc. by invoking translation tools, like a thesaurus, to match the different vocabularies. The workstation should also provide a common interface to all the software tools needed to manipulate and integrate genomic data. Initial developments to approach this vision are being addressed at the Human Genome Center at LBL and at other Genome Centers.

Thus the Human Genome Project challenges not only biology, but also creates interesting new problems for computer science research. Data base designs are challenged by the need for multiple concurrent editing abilities and versions in order to keep an original copy while editing the map. Some kind of alerters are needed to signal the user that some of the information he requests has changed since his last session. The expected high error rate of initial entries may invalidate early maps. The database has to provide easy and fast update methods and error correction capabilities. Plausibility test routines are needed to check and affirm the user input to avoid errors in data entry.

CONCLUSION

Despite all the problems and solutions mentioned above, the analysis of the DNA sequence of the human genome is only part of the long range goal. The information revealed by looking at the DNA sequence will lead to the genes that encode the proteins that run the cellular machinery. The analysis of the proteins invariably leads to problems concerning the structure and ultimately the function of the proteins. Studying the function of proteins returns the emphasis back to genes and their regulation, so now we are addressing just the coversheet of the *Encyclopedia of Man.*

ACKNOWLEDGEMENTS: This work was supported by the Director, Office of Energy Research, Health and Environmental Research, Human Genome Program, of the US Department of Energy under Contract no. DE–AC03–76SF00098 and by a *Schrödinger* fellowship from the Austrian *Fonds zur Förderung der wissenschaftlichen Forschung* to M.Z.

References

[1] Bilofsky HS and Burks C (1988) The GenBank genetic sequence data bank *Nucleic Acids Res* **16**:1861.

[2] Human Gene Mapping 10 (1989) Tenth international workshop on human gene mapping *Cytogenet Cell Genet* **51**:1.

[3] Evans GA and Lewis KA (1989) Physical mapping of complex genomes by cosmid multiplex analysis *Proc Natl Acad Sci USA* **86**:5030.

[4] Olson MV, Hood LE, Cantor CR, and Botstein D (1989) A common language for physical mapping of the human genome *Science* **245**:1434.

[5] Saiki RK, Gelfand DH, Stoffel S, Scharf SJ, Higuchi R, Horn GT, Mullis KB, and Erlich HA (1988) Primer-directed enzymatic amplification of DNA with a thermostabile DNA polymerase *Science* **239**:487.

[6] Ochman H, Gerber AS, and Hartl DL (1988) Genetic applications of an inverse polymerase chain reaction *Genetics* **120**:621.

[7] Triglia T, Peterson MG, and Kemp DJ (1988) A procedure for *in vitro* amplification of DNA segments that lie outside the boundaries of known sequences *Nucleic Acids Res* **16**:8186.

Nomenclature in the Field of Eicosanoids

A.L. Willis, W.L. Smith *, D.L. Smith and R. Eglen
Syntex Research, Palo Alto California;

*Department of Biochemistry,
Michigan State University, East Lansing MI.

Abstract: Eicosanoids are oxygenated derivatives of polyunsaturated fatty acids (PUFAs), generally of 20 carbons in chain length and often with important biological activity; the PUFAs themselves can be named according to a shorthand nomenclature describing chain length, degree of unsaturation and position of the double bond nearest the terminal methyl ("omega") carbon. Arachidonic acid (20:4ω6) is the archetypal precursor for all main classes of the eicosanoids. Prostaglandins (PGs) A-J (ring type) are further classified as to degree of unsaturation in the side chains (eg PGE_1 has one such double bond) and also according to stereochemistry; thromboxanes (A or B, according to ring type) have similar nomenclature.
Leukotrienes (LTs) have a characteristic conjugated triene system and are almost always formed via 5-lipoxygenation. LTA compounds are 5(6) epoxides and LTB compounds are characteristic dihydroxy compounds. LTs C-D are peptidolipids each with a specific amino acid characteristically attached via sulfur to carbon 6 of the fatty acid moiety. Lipoxins (LXs) are characteristic trihydroxy compounds derived by 15-lipoxygenation via an epoxy-15-hydroxy intermediate. LXAs have a 5,6-dihydroxy region and LXBs have a 14,15-dihydroxy region. Both LTs and LXs are (like PGs and TXs) subclassified according to degree of side chain unsaturation. Nomeclature of other oxygenated PUFA metabolites (HETEs etc) is logically adapted from previous usage. Nomenclature of PG-like (prostanoid) receptors are classified into EP, FP, DP and TP types whose characteristic natural agonists are (respectively) E-type PGs, F-type PGs, D-type PGs and thromboxane A_2 but are better defined by response to more specific synthetic prostanoids and newly developed specific antagonists. It is now recognized that the EP receptor may be subclassified into EP_1, EP_2, and EP_3 receptors. Recently, subclassification of peptidoleukotriene receptors has been proposed (PLT_1, PLT_{2a}, PLT_{2b}, PLT_{2c} and PLT_{2d}) and for LTB receptors (LTB_{4Ia} and LTB_{4Ib}. Proposals for more accurate nomenclature for enzymes that form and metabolize eicosanoids are discussed.

K. Loening (Editor)
The Terminology of Biotechnology:
A Multidisciplinary Problem

INTRODUCTION

The term "Eicosanoids" was suggested by E.J. Corey as a general term for oxygenated derivatives of polyunsaturated fatty acids, generally of 20 carbons in chain length and often with important biological activity[1]. Together with their metabolites, there are now several hundred eicosanoids that occur throughout nature, both in mammalian systems and in lower life forms including marine organisms and plants. Biological activity of the eicosanoids in mammalian systems is extremely varied. Indeed, widely differing (even opposing) effects can be produced by small changes in chemical structure. Wide differences in biological actions of individual eicosanoids (especially prostaglandins) are also seen between species and between organs and tissues within species.

In an attempt to more adequately understand these complexities, some logical classifications of eicosanoid structure, biological actions and enzymology have evolved.

An exhaustive description of the nomenclature involved is obviously beyond the scope of this brief chapter, but the reader is referred to more detailed works as appropriate.

NOMENCLATURE OF EICOSANOID STRUCTURES

Structures and biosynthetic pathways of most natural eicosanoids and their metabolites are documented in Volume I of the CRC Handbook of the Eicosanoids(1-4); actions of the numerous synthetic analogs of the prostaglandins are also described (5).

The Essential Fatty acids

Derivation of nomenclature for the prostaglandins (PGs) and their natural polyunsaturated fatty acid precursors is fully described elswhere(1,6,7). Since the latter are derived from dietary sources they are termed the Essential Fatty Acids (EFA) of which several are precursors for the prostaglandins or other eicosanoids. Some polyunsaturated fatty acids (e.g. Mead acid; all cis 5,8,11-eicosatrienoic acid), can also serve as precursors for leukotrienes but are not converted into PGs.

All EFA have the so called "skipped" or "methylene interrupted" pattern of unsaturation and have the general formula:

$$CH_3[CH_2]_k[CH\overset{\underline{cis}}{=}CH.CH_2]_m[CH_2]_n.COOH$$

Arachidonic acid, precursor for most commonly occurring eicosanoids has a total chain length of 20 carbons (4 double bonds) and can be derived biosynthetically from linoleic acid (18 carbons, 2 double bonds) by alternate steps of desaturation and chain elongation at the carboxyl end of the molecule. The final "n" portion of the molecule remains at 6. Consequently, these EFA are of the so called "omega 6" series. A similar "omega-3" series is derived from alpha-linolenic acid, resulting in 5,8,11,14,17-eicosapentaenoic acid, biosynthetic precursor for prostaglandins with 3 double bonds and some lipoxygenase products. Sharing these same enzymatic pathways, oleic acid, derived by desaturation of de novo synthesized stearic acid, is converted into Mead acid that therefore accumulates in tissues of mammals receiving a diet restricted in EFA content.

----

[1]The term "prostanoids" should be restricted only to those compounds related to the prostaglandins. This term includes the thromboxanes (that are similar in structure to PGs) but not leukotrienes and other lipoxygenase products.

Holman(8) proposed a shorthand nomenclature describing chain length, degree of unsaturation and "series" (omega 6, omega 3 etc). Thus arachidonic acid is 20:4ω6, linoleic acid is 18:2ω6, eicosapentaenoic acid is 20:5ω3, and Mead acid is 20:3ω3. IUPAC once proposed the term "n-6" instead of omega-6 (9), but common (even lay) usage has popularized the omega terminology(10).

The Prostaglandins

In the 1930s, "Prostaglandin" was the name given by von Euler(11) to the acidic lipid agent in semen and vesicular gland causing depression of blood pressure, contraction of intestinal smooth muscle and relaxation of uterine smooth muscle. This name (actually a misnomer) arose from the belief that it originated from the prostate. It was not until the 1950s that structural elucidation of the various prostaglandins occurred and a descriptive nomenclature was devised. Systematic nomenclature is based upon a theoretical "prostanoic acid" backbone (Figure 1), and a convenient trivial nomenclature, developed by the Karolinska group, has been universally adopted.

The principal "families" (or "types") of prostaglandin are denoted by the letters E, F etc. These now extend through the alphabet to the letter J and describe both unsaturation of and substituents on the cylopentane ring (Figure 1). Prostacyclin (PGI_2) is a labile bicyclic prostaglandin, with an oxygen bridge between carbons 6 and 9. In addition, prostaglandin nomenclature is further modified by describing degree of unsaturation in the side chain. For example, PGE_1 has one double bond, PGE_2 has 2 double bonds, and PGE_3 has 3 double bonds. The final descriptor of PG nomenclature deal with stereochemistry of the hydroxyl group (especially at the ring). Thus $PGF_{2\text{-alpha}}$ has the hydroxyl groups at positions 9 and 11 below the plane of the other carbons (as conventionally drawn) while $PGF_{2\text{-beta}}$ has the 9 hydroxy group above the plane of the ring.

Less common descriptors in PG nomenclature deal with chain length different from the usual 20 carbons. The term "nor" is used to describe removal of a carbon and so the 18 carbon structure remaining after removal of two carbons by beta oxidation is a donor compound, and the major urinary metabolite of most PGs is a tetranor compound (i.e. shortened by 4 carbons). The term "homo" is used to describe extension by one carbon. Usual extension is of 2 carbons and the terms "dihomo" or "bishomo" are then commonly used. Prostaglandin nomenclature and its historical derivation is fully described in ref.1.

The Thromboxanes

In 1975, Samuelsson's group(12) described a new family of hydroxylated, cyclized polyallelic fatty acids that lacked the prostanoic acid backbone of the classic prostaglandins. These were designated thromboxanes (TX) because of their characteristic oxane ring structure and their probable biological role in aggregating platelets. Their characteristic ring structures are shown also in Figure 1. They are of two types which (following prostaglandin usage) are A and B, the former being extremely unstable but with high biological activity and B being the stable breakdown product that is comparatively inactive.

Punaglandins, Levuglandins and other unusual prostanoids

These compounds from marine organisms and other sources have peculiar (or even opened) ring structures. Suggested abbreviations are PnG, LG etc (10).

Figure 1. Prostanoids

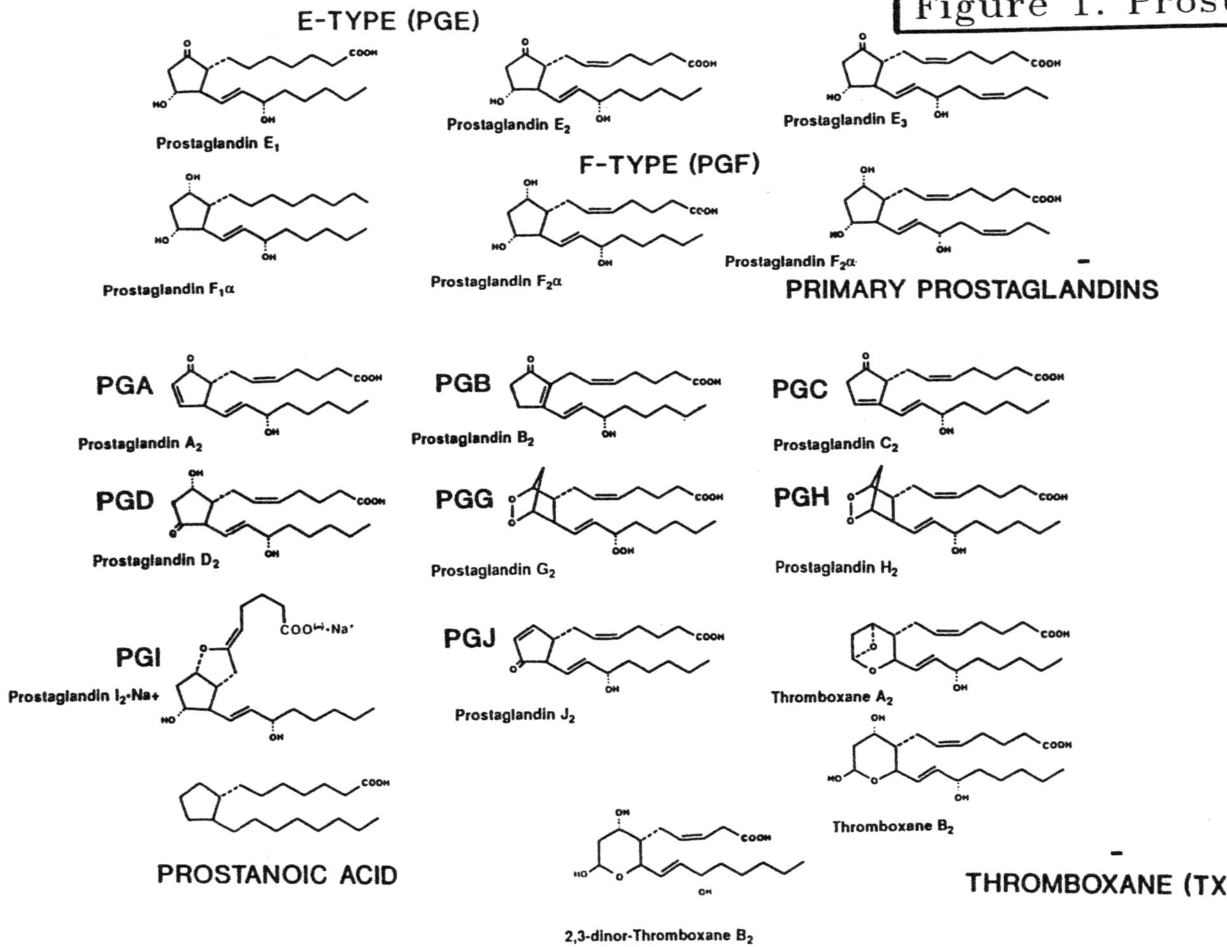

The Leukotrienes

The leukotrienes (LTs) are named for their occurrence and possible role in leukocytes and for their characteristic conjugated triene structure. As before, the main classes or "types" are designated by letter. Thus LTAs are epoxides (generally with the epoxide bridge between the 5 and 6 carbon atoms)[2] while LTBs are dihydroxy oxygenated fatty acids. Leukotrienes C,D,E and F are peptidoleukotrienes. Structures are exemplified in Figure 2 and examples of LT metabolites and leukotrienes derived from polyunsaturated fatty acids other than arachidonic acid are described in detail elsewhere (13,14).

The Lipoxins

The Lipoxins (LXs) are characteristic trihydroxylated products formed via 15-lipoxygenation of arachidonic acid and other 20 carbon EFA. Lipoxin A products have characteristic hydroxylation at carbons 5, 6 and 15, while LXB compounds have 5,14,15 hydroxylation (Figure 3).

Nomenclature for LXA and LXB or their derivatives is as proposed by Serhan et al(15). As for the prostaglandins, numerical subscripts define the number of double bonds in the side chains and the old term "lipoxenes" for products derived from eicosapentaenoic acid is now defunct, being replaced by LXA_5 and LXB_5 (16).

Lipoxygenase products (other than Leukotrienes and Lipoxins).

The first of these to be described was "12-HETE" formed by the 12-lipoxygenation of arachidonic acid by a cytosolic enzyme in platelets(17,18). Later, a plethora of these substances was described, not only derived from arachidonic acid, but also from 20 carbon and 18 carbon polyunsaturated fatty acids both EFA and non EFA.

There was previous confusion in the existing shorthand nomenclature (e.g. EPHETA and HEPA were terms once used to describe the same compound). Smith and Willis(10) proposed a nomenclature that sought to extend and make more rational the most widely accepted usage. With some modification, this terminology was adopted by an Ad Hoc Committee of interested researchers and has become part of the overall nomeclature recomendations going foreward to the International Union of Biochemistry(16). Criteria and guidelines for this nomenclature are summarized in Table 1 and specific working examples are shown in Figures 4 and 5. It will be seen that the only real departure from common usage is use of lower case letters (Hp instead of HP) to define "hydroperoxy" and use of "12-HHTrE to replace the old term "HHT"
The shorthand "HETE" nomenclature used above can be used interchangeably both with the lipoxins and also with the "hepoxilin, trioxilin" nomenclature (Figure 5) proposed by Pace -Asciak(1,19,20) which has not seen wide usage.

NOMENCLATURE OF EICOSANOID RECEPTORS

Characteristic biological actions of synthetic prostanoids (5) and of eicosanoid antagonists(21,22) have been extensively tabulated.

In the area of the prostanoids, there is now some logical classification of putative receptor types, based upon the relative potencies of both natural and synthetic agonists and recently developed specific prostanoid antagonists. The subject of prostanoid receptor classification has been reviewed in great detail by Eglen and

----

[2]The committee on eicosanoid nomenclature recommended that the term leukotriene be restricted to products of 5-lipoxygenation. However, the term leukotriene has also been used in the literature to describe conjugated trienes derived from lipoxygenation at carbons 8 or 15 (ref.2).

Figure 2. Leukotrienes

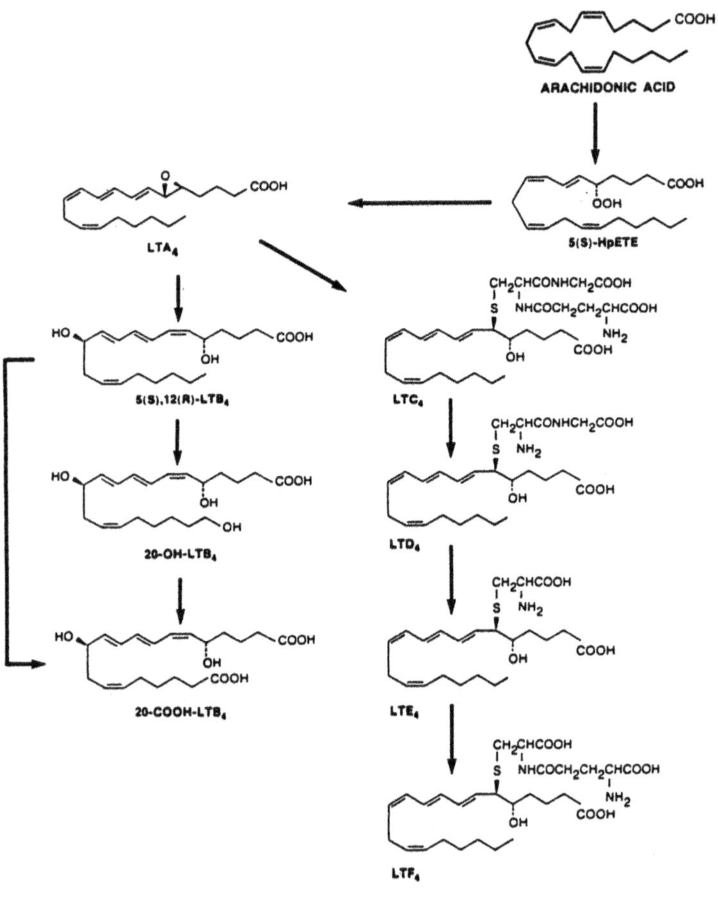

Figure 3. Lipoxins

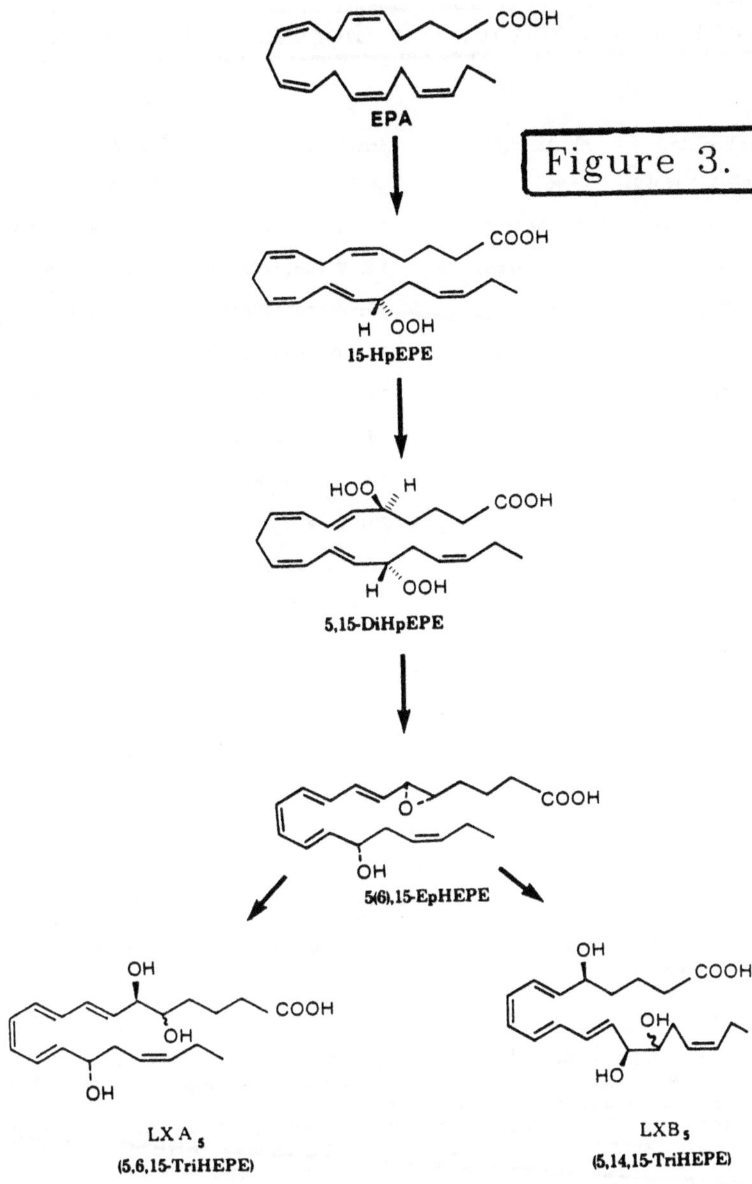

Previously termed Lipoxenes

Table I Guidelines for Shorthand Nomenclature

Number of Substituents	Substituent Name	Number of Carbons	Number of Double Bonds
1: no name	Hydroxy: H	12: D (dodeca)	1: ME (monoenoic)
2: Di	Hydroperoxy: Hp	14: T (tetradeca)	2: DE (dienoic)
3: Tri	Epoxy: Ep	15: P (pentadeca)	3: TrE (trienoic)
4: Tetra	Keto: Oxo	16: Hx (hexadeca)	4: TE (tetraenoic)
5: Penta		17: H (heptadeca)	5: PE (pentaenoic)
6: Hexa		18: O (octadeca)	6: HE (hexaenoic)
		19: N (nonadeca)	
		20: E (eicosa)	
		22: Do (docosa)	

Position of Substituent Groups

The position of the substituent is indicated by a number, counting the carboxy terminal carbon as 1. For epoxides, the numbers should use parentheses for the second number [e.g. 5(6)-EpETE]. For hydroxyl, hydroperoxyl and oxo groups, the number should take the usual form [e.g. 5-HETE].

Double Bond Position

The positions of the double bonds should be indicated with a number to designate the location of carbon nearest the carboxyl end of the molecule and a letter, either "Z" or "E", to indicate the **cis** and **trans** geometry, respectively.
The designation for position of the double bond should be in parentheses at the beginning of the shorthand name. For example, 5-HETE should be (6E,8Z,11Z,14Z)5-HETE. The designation for the double bond position need only be used in the abbreviations section of a manuscript, unless it is appropriate to use this designation throughout the manuscript for the purpose of clarity (from ref.16).

Figure 4. Examples of Shorthand Nomenclature

			Substituent prefix		Chain length		Suffix: Number of unsaturations		
Precursor	Compound	Position number prefix	Symbol	Description	Symbol	Description[a]	Symbol	Description	Comments
Arachidonic acid (20:4ω6)	5-HETE	5-	H	Hydroxy	E	Eicosa (20)	TE	Tetraenoic acid	
	12-HHTrE	12-	H	Hydroxy	H	Heptadeca (17)	TrE	Trienoic acid	Cyclo-oxygenase by-product
	15-HpETE	15-	Hp	Hydroperoxy	E	Eicosa (20)	TE	Tetraenoic acid	
	5(6)-EpETrE	5(6)-	Ep	Epoxy	E	Eicosa (20)	TrE	Trienoic acid	
	12(S)-HETE	12(S)-	H	Hydroxy	E	Eicosa (20)	TE	Tetraenoic acid	(S) and (R) denote sterochemistry of hydroxy group (see ref. 2)
	12(R)-HETE	12(R)-	H	Hydroxy	E	Eicosa (20)	TE	Tetraenoic acid	
	5,15-DiHETE	5,15-	DiH	Dihdroxy	E	Eicosa (20)	TE	Tetraenoic acid	
	13,14(15)-HEpETrE	13,14(15)-	HEp	Hydroxyepoxy [13] [14(15)]	E	Eicosa (20)	TrE	Trienoic acid	
	5(6),15-EpHETE	5(6),15-	EpH	Epoxyhydroxy [5(6)] [15]	E	Eicosa (20)	TE	Tetraenoic acid	
	14-Oxo-TTE	14-	Oxo-	Oxo (aldehyde)	T	Tetradeca (14)	TE	Tetraenoic acid	
	12-Oxo-DTrE	12-	Oxo-	Oxo (aldehyde)	D	Dodeca (12)	TrE	Trienoic acid	
Dihomo-γ-linolenic acid (20:3ω6)	(5Z,8E,10Z)12-HETrE	12-	H	Hydroxy	E	Eicosa (20)	TrE	Trienoic acid	
	8,9,12-TriHEDE	8,9,12-	TriH	Trihydroxy	E	Eicosa (20)	DE	Dienoic acid	
	8,15-DiHpETrE	8,15-	DiHp	Dihydroperoxy	E	Eicosa (20	TrE	Trienoic acid	
	12-HHDE	12-	H	Hydroxy	H	Heptadeca (17)	DE	Dienoic acid	Cyclo-oxygenase by-product
	10,11(12)-HEpHME	10,11(12)-	HEp	Hydroxyepoxy [10] [11(12)]	H	Heptadeca (17)	ME	Monoenoic acid	Possible cyclo-oxygenase product

[a]Number of carbons is in parentheses.

Figure 5. Further Nomenclature Examples

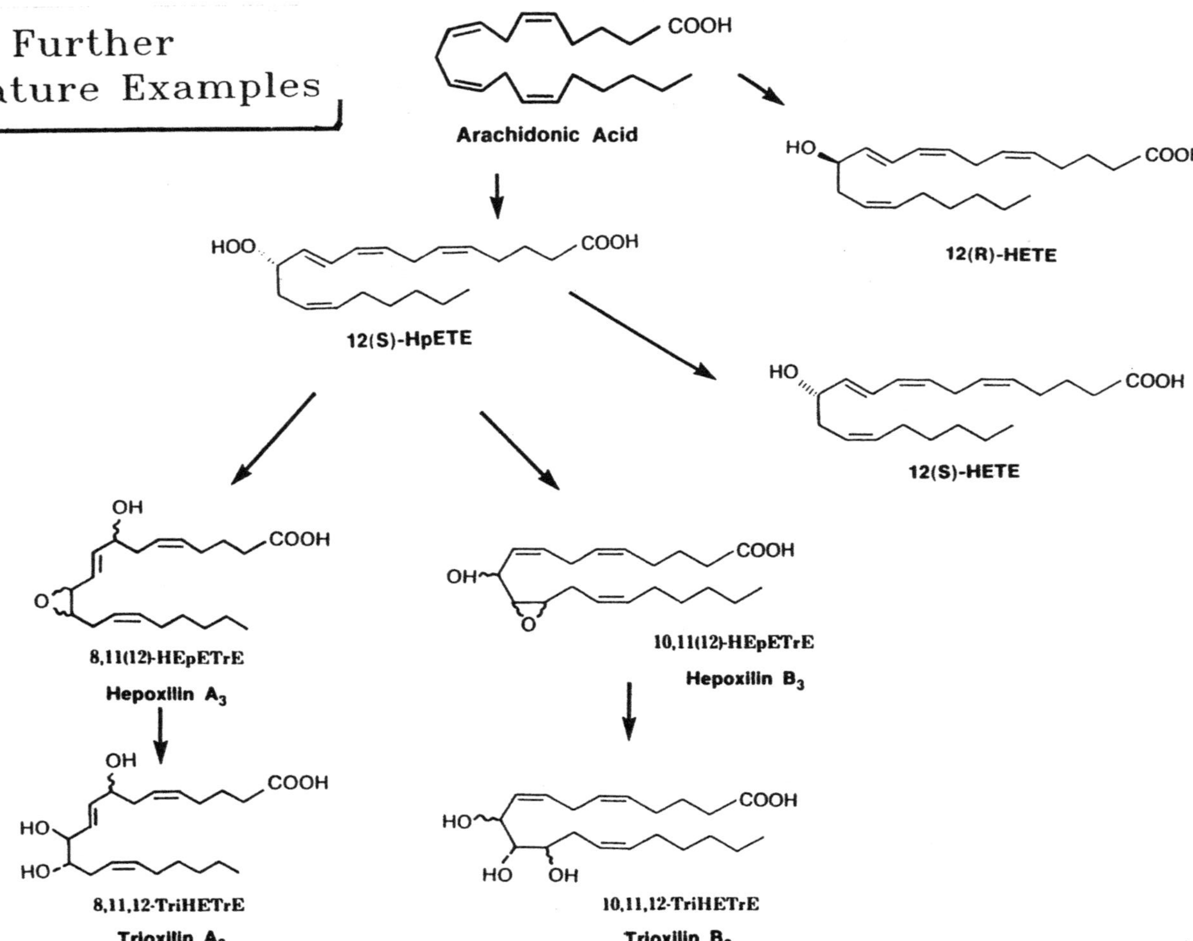

Whiting(23), who not only reviewed present state of the art, but also earlier attempts at receptor classification. The general classification of prostaglandin receptors proposed by the Glaxo group of Kennedy, Coleman and co-workers(24-27) is now generally accepted and is being extended as new synthetic prostanoid antagonists are developed.

Using the criteria of Furchgott(28) biological basis for the existence of receptors and their subtypes can be established. It is generally accepted that receptor analysis should be first assessed in vitro, since under such conditions, concentrations of agonist and antagonist at the receptor can be reasonably assessed(23). Complications in vivo include metabolism of prostanoids to metabolites that could have different effects and release of endogenous substances (including eicosanoids) with additional biological activity.

Ideally, one hopes to eventually show correlation of biological activity with binding of the agonist to the proposed receptor and its appropriate prevention by specific antagonists. Eventually, isolation and structural elucidation of the receptor and definition of "second messengers" would be desirable. In the meantime, postulation of prostanoid receptor-mediated agonist and antagonist activities can give a hypothetical basis for drug development studies. The success of this approach in other areas is well documented (29,30).

According to the Glaxo scheme, there are receptors for PGE_1 and PGE_2 (EP receptors); $PGF_{2\text{-alpha}}$ (FP receptors); PGD_2 (DP receptors); thromboxane A_2 or synthetic PGH_2 analogs (TP receptors) and prostacyclin and its analogs (IP receptors).

EP Receptors

This type of receptor, for which the natural agonists are PGE_1 and PGE_2, produce contractions of guinea-pig trachea and gastrointestinal muscle of guinea-pig, dog and other species. Relaxation of precontracted cat and guinea-pig trachea are also mediated by EP receptors(23,25,26). On the basis of selective antagonism, there are now 3 subclassifications of EP receptor (EP_1, EP_2, and EP_3)[3] EP_1 receptors are stimulated most selectively by 16,16-dimethyl PGE_2 (which also has TP agonist action in appropriate systems), while EP_2 receptors may be selectively stimulated by AY 23626, eliciting mainly relaxant effects. EP_3 receptors are now defined as as mediating reduced acid secretion in the rat gastric mucosa and reducing electrically stimulated twitch in the guinea-pig vas deferens. Sulprostone is the best example of an agent acting to stimulate EP_3 receptors although it also acts at EP_1 receptors (27,31); similar effects are produced by RS20216(32). EP_1 and EP_3 receptors are distinguished on the basis of antagonism of the former by SC19220 or AH6809 (27,31).

FP Receptors

The cat and dog iris muscle (but not bovine iris muscle) are highly sensitive to the actions of $PGF_{2\text{-alpha}}$. Of other prostanoid types, PGE_2 and prostacyclin (PGI_2) have negligable activity in this system, but PGD_2 has some activity, with lesser effects being produced by the endoperoxide analog (and thromboxane agonist) U-46619. On the basis of these facts, it has been proposed by Coleman et al(23,26) that these two tissues possess a distinct FP receptor. The synthetic prostanoids cloprostenol and fluprostenol also appear to act as selective FP agonists(26). Since, however, no specific antagonist for the FP receptor has been described, actual existence and possible sub-types of FP receptors must still be regarded as a working hypothesis.

----

[3]It has been speculated that EP_1 receptors are likely coupled to Ca^{++} mobilization, EP_2 to stimulation of adenylate cyclase and EP_3 to inhibition of adenylate cyclase.

DP Receptors

The concept of the DP receptor (whose natural agonist is PGD_2) had until recently been the most poorly defined, since no preparation had been described that contained DP receptors alone and there was lack of specific DP antagonists. However, a high affinity specific antagonist (BW868C) has been described(33), which antagonizes the actions of PGD_2 on platelets and vasculature.

IP Receptors

Prostacyclin (PGI_2) is the natural agonist for this proposed receptor. Inhibition of aggregation of rat platelets (which are unresponsive to PGD_2) was used to delineate this receptor, prior to the discovery of PGD antagonists (which may now be used on human or other platelet preparations responsive to PGD_2). The synthetic PGI_2 analog iloprost has been used to study this receptor, since it appears to possess selective IP agonist properties and (unlike PGI_2) is chemically stable. On platelets, PGE_1 appears to share the same IP receptor as PGI_2 (34)although activity of these two prostanoids largely differ in virtually every other system (1). No selective IP receptor antagonist has been defined.

TP Receptors

This receptor type is thought to mediate contractions of a range of vascular smooth muscle preparations by U46619 (the thromboxane A_2 mimetic). Much less effect is produced by PGD_2 (thought to have some TP agonist activity) with negligable activity produced by other prostanoid agonists (PGI_2; $PGF_{2\text{-alpha}}$ and PGE_2).

There are now a number of potent and selective TP antagonists available, some of which have clinical potential in the treatment of ischemic vascular disease(35), especially when also possessing inhibitory activity against thromboxane A synthase (35,36).

Leukotriene (LT) Receptors

Although two distinct binding sites (cell membrane and subcellular) were described for the peptidoleukotrienes, only recently has functional subdivision of leukotriene receptors been described. This classification by Krell et al(37) describes subdivision of both peptidoleukotriene and non-peptide (LTB_4) receptors.

Contractions of the isolated guinea-pig trachea may be considered to be mediated via 5 distinct peptidoleukotriene receptors: PLT_1 receptors are activated by LTC_4, the contractions of which are **not** blocked by available leukotriene antagonists.

A Second major subtype (PLT_2) of putative leukotriene receptors on the guinea-pig trachea mediates contractions that are induced by LTD_4 and E_4 (not C_4) and can be blocked by the classic antagonist FPL55712. However subdivisions of this receptor seem to exist based upon affinity for antagonists. Thus, PLT_{2a} receptors on the guinea-pig trachea are strongly inhibited by FPL55712 (and a variety of other antagonists) that therefore can be considered to have a high affinity for this receptor(37). By contrast, the putative receptor subtype PLT_{2b} has low affinity for FPL55712 (which therefore weakly inhibits contractions)(37). Two other subtypes of the PLT_2 receptors have also been postulated:
Human isolated bronchial smooth muscle is induced to contract by the peptidoleukotrienes in general (C_4, D_4 and E_4) via a proposed PLT_{2c} receptor. This proposed receptor is apparently blocked effectively by FPL55712 and other antagonists (SKF104353, ICI204219, ICI204219 and ICI198615) and is clearly different from those on the guinea-pig trachea(37).

Finally, isolated guinea-pig ileum is strongly induced to contract by LTD_4 and E_4 and the contractions are inhibited only by FPL55712 and LY17188. The putative receptor involved therefore seems distinct and has been termed PLT_{2d} (37).

For the actions of the non-peptide leukotrienes, receptor subdivisions have not been clearly defined. However in polymorphonuclear (PMN) leukocytes, there appear to be both high affinity and low affinity receptor sites for LTB_4 (38). The high affinity receptor has been given the suggested name LTB_{4Ia} and seems associated with the effects of LTB_4 in increasing adherence and motility of the leukocytes(37). By contrast, the low affinity receptor (LTB_{4Ib}) seems associated with release of lysosomal enzymes and bursts of oxidative activity (37).

As a precautionary note, the epithelial layer of guinea-pig trachea seems to modulates contractile effects of LTE_4, via release of 5-lipoxygenation products (39). Thus, even simple in vitro test systems can give misleading results, since more than one cell type is involved and interconversion of leukotrienes or release of endogenous substances may still occur.

ENZYMES INVOLVED IN EICOSANOID FORMATION AND METABOLIC INACTIVATION

It is not intended here to exhaustively define enzymology of the eicosanoid system and its classic(1,40) or newly recommended terminology which is fully described elsewhere(16). Apart from the "basal" synthesis of prostaglandins (and perhaps other eicosanoids) that seems to result from metabolic pools of precursor (involved in interconversions of EFA), it is generally thought that the rate limiting factor for eicosanoid production is availability of free arachidonic acid (or related polyunsaturated precursors) liberated by hydrolytic enzymes(1,41). These enzymes can include phospholipases, the release of free fatty acids from triglycerides during lipolysis or action of lipoprotein lipase, and (in the adrenal cortex) cleavage of cholesteryl esters(1,41).

However the enzymology that is most relevant to the subject matter of this article involves formation of the eicosanoids from these precursors and their subsequent interconversions and metabolic inactivation. Some key examples are given below:

Prostaglandin endoperoxide synthase

Prostaglandins are formed via cyclo-oxygenation of arachidonic acid and similar polyunsaturated precursors by the action of an enzyme commonly termed cyclooxygenase, and once called PG-synthetase. This enzyme has the newly recommended name of "prostaglandin endoperoxide synthase"(16) since formation of the endoperoxides (PGs G and H) is the main reaction. However, it is now recognized that this enzyme may co-oxygenate other substrates, for instance conversion of columbinic acid into 9-hydroxyoctadecatrienoic acid(42) and conversion of Mead acid to 13-hydoxy-5,8,11-eicosatrienoic acid (43).

Prostaglandin D,E, and I synthases

These enzymes convert the PG endoperoxides into each prostaglandin type. These recommended terms (16) replace those of (respectively) PGH-PGD isomerase, PGH-PGE isomerase (endoperoxide isomerase), and prostacylin synthase.

Prostaglandin F synthase

This term is now recommended(16) for the enzyme that converts PGD to PGF compounds and also converts PG endoperoxides to PGF compounds. This terminology therefore replaces previous "11-ketoreductase" and "reductase" usage.

Enzymes that form 15-keto derivatives of PGs

There are now three distinct enzymes recognized(16):
15-hyroxyprostaglandin dehydrogenase acts on PGs E_2, $F_{2\text{-alpha}}$, and B_1, but not upon PGD (or PGI) compounds.

By contrast, 15-hydroxyprostaglandin D dehydrogenase ($NADP^+$) is specific for PGD_2 (and presumably other PGDs), while 15-hydroxyprostaglandin I dehydrogenase ($NADP^+$) is specific for prostaglandin I_2 (although it may well act upon PGI_3).

Thromboxane A synthase

This enzyme converts PG endoperoxides to thromboxane A compounds and (in some unexplained way) catalyzes the formation of malondialdehyde (MDA) that is also formed during spontaneous decomposition of the endoperoxides to 12-HHTrE (formerly HHT). Thus MDA formation has often been used as an indirect marker of both thromboxane and PG synthesis in platelets (1,44).

Lipoxygenases

It is, of course, impossible to review the complexities of the lipoxygenase area and its nomenclature in the space of one or two paragraphs.

However, it is worth noting that arachidonate 12-lipoxygenase acts on several polyunsatured fatty acids, including 22:6ω3(18) with structural specificity such that the term "omega 9 lipoxygenase" was once raised as a possibilility(41).

Similarly, arachidonate 15-lipoxygenase is of two types. Type I occurs in plants and Type II occurs in mammalian cells. Since at least one source of the mammalian enzyme converts linoleic acid (18:2ω6) to 13-HODE (45), omega 6 lipoxygenase was once suggested as a more appropriate term (41,45).

CLOSING REMARKS

Like any language, scientific terminology is limited in precision and sometimes the most strictly logical terminology is too clumsy for day to day use, especially in verbal communication. The natural course of events then leads to attachment to less logical terminology that we have"become used to". Examples are the use of "prostaglandins" (which do not in fact derive mainly from the prostate) and "eicosanoids" (many of which do **not** have twenty carbon atoms).

There is currently only one example where a previous decision with many ramifications needs to be reversed:

This is the previous IUPAC decision (46) to change the spelling of "eicosanoids" to "icosanoids" (as an offshoot from a decision to anglicize polyunsaturated fatty acid nomenlature). Unless a change is made to revert to the original Greek-derived spelling, all "HETEs" would have to be renamed " HITEs" (10).

With this minor exception, eicosanoid nomenclature has evolved in a logical and pragmatic fashion as a consequence of adequate dialog between investigators.

REFERENCES

1. Willis AL (1987) In: Willis AL (ed) Handbook of the Eicosanoids Volume IA, p3, CRC Press, Boca Raton FL.

2. Smith DL (1987) In: Willis AL (ed) Handbook of the Eicosanoids Volume IA, p47, CRC Press, Boca Raton FL.

3. Jackson Roberts,II L (1987) In: Willis AL (ed) Handbook of the Eicosanoids Volume IA, p233, CRC Press, Boca Raton FL.

4. Smith DL, Stone KJ, Willis AL (1987) In: Willis AL (ed) Handbook of the Eicosanoids Volume IA, p245, CRC Press, Boca Raton FL.

5. Muchowski JM (1987) In: Willis AL (ed) Handbook of the Eicosanoids Volume IB, p19, CRC Press, Boca Raton FL.

6. Mead JF, Willis AL (1987) In: Willis AL (ed) Handbook of the Eicosanoids Volume IA, p85, CRC Press, Boca Raton FL.

7. Brenner RR (1987) In: Willis AL (ed) Handbook of the Eicosanoids Volume IA, p99, CRC Press, Boca Raton FL.

8. Holman RT (1966) Progr. Chem. Fats and other Lipids 9: 3

9. IUPAC-IUB Commission on Biochemical Nomenclature (1967) J Biol Chem 242: 4845

10. Smith DL, Willis AL (1987) Lipids 22: 983

11. Von Euler US (1982) Progr Lipid Res 20: xxi

12. Hamberg M, Svensson J, Samuelsson B. (1975) Proc Natl Acad Sci USA 72:2994

13. Borgeat P (1987) In: Willis AL (ed) Handbook of the Eicosanoids Volume IA, p193, CRC Press, Boca Raton FL.

14. Hammarstrom S (1983) Ann Rev Biochem 52: 355

15. Serhan CN, Wong PY-K, Samuelsson B (1987) Prostaglandins 34: 201

16. Smith WL Borgeat P, Hamberg M, Jackson Roberts II L, Willis AL, Yamamoto S, Ramwell PW, Rokash J, Samuelsson B, Corey EJ, Pace-Asciak CR (1990) In: Methods in Enzymology, Academic Press, New York, N.Y. In press.

17. Hamberg M and Samuelsson B (1974) Proc Natl Acad Sci USA 71: 3400

18. Nugteren DH (1977) In: Silver MJ, Smith JB, Kocsis, JJ (eds) Prostaglandins in Hematology, p 11, Spectrum Publications, Jamaica N.Y.

19. Pace Asciak CR, Granstrom Samuelsson B (1983) J Biol Chem 258: 6835

20. Pace-Asciak CR, Martin JM (1984) Prostaglandins Leukotrienes Med. 16: 173.

21. Sanner JD (1989) In: Willis AL (ed) Handbook of the Eicosanoids Volume II, p285, CRC Press, Boca Raton FL, 285.

22. Musser JH, Kreft AF, Lewis AJ. In: Willis AL (ed) Handbook of the Eicosanoids Volume II, p345, CRC Press, Boca Raton FL.

23. Eglen RM, Whiting RL (1989) In: Willis AL (ed) Handbook of the Eicosanoids Volume II, p273, CRC Press, Boca Raton FL.

24. Kennedy L, Coleman RA, Humphrey PPA, Levy GP, Lumley P (1982) Prostaglandins 24: 667.

25. Coleman RA, Humphrey PPA, Kennedy F, Lumley P (1984) Trends Pharm Sci 5: 303

26. Coleman RA Humphrey PPA, Kennedy I (1985) In: Kalsner S (ed) Trends in Autonomic Pharmacology Vol 3,Chapter 3, Taylor and Francis, London

27. Coleman RA, Kennedy I, Sheldrick, RLG, Tolowinska IY (1988) Br J Pharmacol 95: 407.

28. Furchgott RF (1972) In: Blashco H, Muscholl E. (eds) Handbook of Experimental Pharmacology, Catecholamines, Volume 33: 283.

29. Green JP (and Supplemental Commendium) (1990) Trends Pharmacol Sci 11: 13.

30. Cytokines Centerfold (1989) Trends Pharmacol Sci 10:

31. Reeves JJ, Bunce KT, Sheldrick RLG, Stabels R (1988) Br. J. Pharmacol. 95: 805

32. Waterbury DL, Faurot GF, Eglen RM (1990) Invest Opthal Vis Sci 31: 233

33. Giles H, Leff P, Bolofo, ML, Kelly MG, Robertson AD (1988) Br J Pharmacol 95: 522.

34. Siegl AM, Smith JB, Silver MJ, Nicolaou KC, Ahern D (1978) J Clin Invest, 63: 215.

35. Willis AL, Smith, DL (1989) Eicosanoids 2: 69

36. De Clerck F, Beetens J, de Chaffoy de Courcelles D, Vercammen E, Freyne E, Janssen PAJ (1989) In: Prostaglandins in Clinical Research: Cardiovascular System. Schror K, Sinzinger H (eds), p567, Alan R Liss, New York NY.

37. Krell RD, Aharony D, Buckner CK, Kusner EJ (1990) In: Advances in Prostaglandin, Thromboxane and Leukotriene Research, Raven Press, New York, NY. In Press.

38. Lin AH, Ruppel PL, Gorman RR (1984) Prostaglandins 29: 837.

39. Buckner CK, Fedyna JS, Robertson JL, Will JA, England DM, Krell RD, Saban R (1990) J. Pharmacol Exp Ther 252: 77

40. Nomenclature Committee of IUB (1984) Enzyme Nomenclature, Academic Press, New York, NY.

41. Willis AL In: Willis AL (ed) Handbook of the Eicosanoids Volume II, p3. CRC Press, Boca Raton FL.

42. Elliott WJ, Morrison AR, Sprecher H, Needleman P (1985) J. Biol Chem 260: 987

43. Elliott WJ, Morrison AR, Sprecher H, Needleman, P. (1986) J. Biol Chem 261: 6719

44. Smith JB, Ingerman CM, Silver MJ (1976) J. Lab. Clin. Med 88: 167

45. Soberman RJ, Harper TW, Betteridge D, Lewis RA, Austen KF (1985) J Biol Chem 260: 4508

46. IUPAC Commission on Nomenclature of Organic Chemistry (1979) In: Rigaudy J, Klesney SP (eds) Nomenclature of Organic Chemistry: Sections A,B,C,D,E,F, and H. Pergamon Press, Oxford, England, p.5

United States Adopted Names for Drugs

Donald O. Schiffman, Ph.D.

American Medical Association
535 North Dearborn Street
Chicago, IL 60610

Abstract: The United States Adopted Names (USAN) Council is the United States committee responsible for negotiating nonproprietary names for the pharmaceutic industry. It is a non-governmental organization sponsored by the American Medical Association, the United States Pharmacopeial Convention, and the American Pharmaceutical Association. Nonproprietary names serve a useful function that cannot be supplanted by the chemical name or trademark. The goal of the Council is to produce simple, short, unique, and useful names that can be used without restriction in the public domain. The Council also has close liaison with the International Nonproprietary Names (INN) Committee of the World Health Organization and works to establish Council-generated names as international names.

INTRODUCTION

A name has substance and can be an item of great integrity. It becomes the "descriptor" of a person or product. If a pharmaceutical manufacturer develops a useful drug and a less than outstanding name is coined for it, the compound will overcome this defect and its name will become a hallmark of good therapy. Conversely, a name that everyone accepts as being well constructed, attractive, and appropriate will not salvage a poor drug.

A BRIEF HISTORY OF DRUG NOMENCLATURE IN THE UNITED STATES

Nonproprietary names serve a useful function, particularly if they are short, simple, and easy to pronounce. However, the United States Adopted Names (USAN) Council and other national and international nomenclature groups were not the first to recognize this concept. Before most of these nomenclature bodies existed, pharmaceutical manufacturers and private research groups were coining nonproprietary names.

The inherent "goodness" or "badness" of the early names was not the force that brought the national and international groups into existence, because some of the early nonproprietary names were very good. What actually stimulated the movement to create committees to develop names for use in thepublic domain was the one factor that none of the early names held in common: A lack of national, or better still, global uniformity in the way names were created.

Thus, in the 1950s and 1960s, nomenclature committees began to work in concert to develop a standardized and logical system that would apply to the creation of nonproprietary names worldwide. It is and has been an evolving process as the groups learned from their mistakes, successes, and each

K. Loening (Editor)
The Terminology of Biotechnology:
A Multidisciplinary Problem

other. Also, the liaison between the committees provide a mechanism for understanding common problems despite some differences in the basic philosophy of creating the nonproprietary names.

In the United States, the function of developing nonproprietary names was first undertaken by those organizations responsible for publishing the *United States Pharmacopeia* and the *National Formulary.* During the nineteenth century, both individual drugs and mixtures were listed in these compendia. As greater numbers of single-entity compounds began to be developed and marketed, a more timely mechanism was needed to create nonproprietary names than that provided during the 10-year periods between revisions of the USP and NF. In 1910, the Council on Pharmacy and Chemistry of the American Medical Association began to assign nonproprietary names to new chemical substances used as drugs.

Like other areas of business and science, the drug industry matured in the twentieth century. As it became increasingly skilled in the development of specific products for treating specific diseases, an unprecedented number of drugs began to proliferate, especially following World War II. By the early 1960s the mechanisms in use for coining names needed to be revised in order to develop nonproprietary names that were short, that conveyed useful information to physicians and other individuals using the names, and that could be used globally regardless of language barriers. In 1961, the United States Pharmacopeial Convention joined the American Medical Association to form the AMA-USP Nomenclature Committee. The primary objective of this new committee was to develop single, suitable, nonproprietary names for each new drug introduced into the U.S. marketplace. The term "United States Adopted Name" or USAN was used to designate the nonproprietary name. This title and acronym were carefully chosen to indicate that the committee had no specific sanction from the federal government. The AMA-USP Committee was joined by the American Pharmaceutical Association in 1964 and became known as the USAN Council. In 1967, the U.S. Food and Drug Administration began to participation in the deliberations of the USAN Council through appointment of a liaison member to the Council. The FDA strengthened this relationship in 1984 by announcing that it would discontinue adding to its official list of drug names and use USAN Council-derived names as established names for labeling and advertising purposes. The USAN Council is unique in that it is the only privately sponsored organization of its kind in the world.

THE NONPROPRIETARY NAME

Why is there a need for nonproprietary names? Why is the trademark not sufficient identification for a drug? It has been argued that there are already two and, in some instances, three names applied to each therapeutic compound, so why create yet another?

It is obvious that the precise chemical name derived from the chemical structure of a compound often is long, complex, and unwieldy. For example, it would be difficult, if not impossible, to refer to such a name repeatedly in a scientific manuscript. At the other end of the spectrum, trivial names that may be short and easy to utilize also may be imprecise and uninformative. Trademarks and manufacturer code designations could be used as primary identifiers, but this type of terminology is proprietary and cannot provide the freedom and impunity afforded a name in the public domain.

The nonproprietary name protects the trademark. This was best explained by Windsor Cutting, a past chairman of the USAN Council, when he wrote an editorial entitled *A Note on Names* that appeared in the March 1963 edition of the journal of *Clinical Pharmacology and Therapeutics*. He stated :

> "From the standpoint of the pharmaceutical industry a nonproprietary or public name is necessary to protect the trademark. If at any time it is shown that the trademark has been so commonly used as to become, in truth, the common name, the trademark is lost. The common name "aspirin" was lost as a trademark in the United States as a result of its general use."

There also are other reasons for the existence of nonproprietary names. Although the average physician never reads the *United States Pharmacopeia and the National Formulary* and has probably never heard of the *USAN and the USP Dictionary of Drug Names*, these books are of great value to the health care professionals by their mere existence. This fact assures them that there are agencies that monitor standards and indications for drugs. They therefore are freed from concern about the general surveillance of such matters.

The agencies that publish these books of reference use only, or principally, nonproprietary names. They are at least quasi-public organizations and strictly refrain from partisanship. Using the public domain name not only relieves them of any favoritism, but also removes the dilemma presented by multiple trademarks for a single drug. Similarly, teachers, writers of textbooks, and editors of journals, whenever possible, choose to avoid bias by using the public name.

Another value inherent in a USAN is that it almost always has international currency. The World Health Organization (WHO) name and the name in the British Pharmacopoeia, for example, will ordinarily be the same as the U.S. name. During the process of selection, the tentatively accepted USANs are submitted to international agencies to be sure that they are acceptable and that conflicts do not exist. Thus, a real value of a nonproprietary name is to have a common name in the public domain that identifies each drug distinctively so that an individual interested in drug nomenclature can identify the product regardless of where in the world one lives or what language one speaks. This is a goal unattainable with a trademark, since the same drug can have many trademarks.

THE IMPORTANCE OF GLOBAL NAMES

A few countries, including the United States, have national drug nomenclature committees. These bodies develop names for use by their respective countries. It is obvious that to have each country coining nonproprietary names independent of each other is not the ideal situation. In 1951 the WHO created the International Nonproprietary Name (INN) Committee to coin names for international use. It was assisted by the existing national committees and by the drug industry in those countries that did not have a national committee. Thereby began the process of international negotiations that still exists today. The system is relatively effective but does not always proceed smoothly. Often considerable compromise must be made by a national committee, a drug manufacturer, or the INN Committee itself; however, the end benefit of having universal names for use throughout the world is worth the effort. Nevertheless, despite the close liaison between the various nomenclature committees, achieving global names is not always realized. While some national committees are bound by agreement with the WHO to accept names created by the INN Committee, others are not. The USAN Council is an example of the latter. On rare occasions there are valid reasons for a divergence of names.

Since its inception, the USAN Council has worked closely with the INN Committee and, recently, with other national nomenclature groups such as the British Pharmacopeia Commission, the French Pharmacopeia Commission, and the Japanese Accepted Names Committee. Because most major pharmaceutical companies are multinational, the need for global names is greater than ever before.

THE DEVELOPMENT OF A USAN

The USAN staff has created a submission form that is used to apply for a USAN. A copy of this form can be found in the current edition of the *USAN and the USP Dictionary of Drug Names*, a soft-covered book that can be purchased from the United States Pharmacopeial Convention, Inc., and in the second edition of the *USAN Handbook*, a booklet available free of charge from the USAN Secretariat at the American Medical Association. These forms may be freely photocopied. Individual forms may be obtained by writing or calling the USAN offices at the American Medical Association.

From the point of view of the USAN Council, the ideal time for a drug manufacturer to make a submission for a nonproprietary name is after there is some human experience with the experimental drug. The Council believes that if a candidate drug has progressed to the point where even limited clinical trials have been initiated, the manufacturer developing the compound is optimistic that the substance will become a marketable drug.

When making an application, certain types of information are needed by the Council before it will consider a name for adoption:

Suggested Names

An applicant should propose one or two nonproprietary names. For a name to be useful, it must have two ideal qualities: It should be short and should tell a story. A suggested name must incorporate an established family name where appropriate. An example is the "-olol" stem used for drugs that have beta blocking action and also incorporate a particular chemical chain.

Chemical Names or Descriptions

The USAN Council uses Chemical Abstracts Service (CAS) as its authority for chemical nomenclature. If the manufacturer has already established a CAS name for its compound, this name should be listed on the submission form. If not, then the USAN Secretariat will obtain one.

Structural Formula

Without a structural formula it may be impossible for the Council to determine what family stem should be used for naming a particular drug and whether a USAN or INN may already exist for the compound.

Molecular Formula

A one-line molecular formula should be supplied.

Molecular Weight

A molecular weight should be supplied.

Trivial Name or Names

Sometimes, a new compound will acquire a trivial name long before a nonproprietary name or trademark has been created. Often this name originates in the laboratory or in the lay press and, occasionally it will creep into the scientific literature. The Council wants to be made aware of such names but requests manufacturers not to create, use, or in any way encourage the development of trivial names for new drugs. The fact that a trivial name has become entrenched in the literature will not ensure its adoption as a nonproprietary name and may only add confusion when a proper nonproprietary name is constructed.

Other Items

The CAS registry number should be supplied or, if no number has been assigned, the Secretariat will obtain one. Company code designations assigned to a candidate drug should be made available, particularly if the code has been used in published scientific papers. If a trademark has been issued for the drug, it should be provided. Principal therapeutic uses and pharmacologic actions should be indicated, since this information may influence the stem name selected.

There are three additional concepts that need to be considered in the submission and creation of a nonproprietary name, but these are not really part of the application guidelines. The first concept deals with the USAN in the public domain. There is an unfortunate tendency among drug manufacturers to register trademarks that are extremely similar to USANs, thereby clearly infringing on previously established names. Not infrequently, a major portion of the USAN, often the stem, is incorporated into the trademark. This practice is a disservice to those who are trying to systematize names in a meaningful and useful nomenclature program. It is particularly deplorable when a firm uses a stem created by the USAN Council or the INN Committee and later protests that a subsequent nonproprietary name adopted for a related compound is using the same stem and the firm now considers the new name to be in conflict with its trademark.

The second concept deals with the fee-for-service. As one of the original sponsors of the USAN Council, the AMA agreed to support the Secretariat of the Council at no cost to the drug industry or the government, a commitment it honored for a quarter of a century. During that time approximately 2,500 new nonproprietary names were created by the Council. Also, working with other national and international nomenclature committees, the Council has been instrumental in having the majority of USANs accepted as global names.

Because the cost of assigning nonproprietary names has escalated greatly over this same quarter of a century, the AMA made a decision that it could no longer bear the total burden of all USAN Secretariat activities. On November 15, 1985, the AMA began assessing a nominal user's fee for each USAN negotiation.

The third concept is concerned with the cooperative effort of the manufacturer. Although USANs are in the public domain, the USAN Council clearly understands that a manufacturer still has considerable interest in what nonproprietary name is coined for its drug. The Council will not adopt or publish a name that has not been agreed to and accepted by the manufacturer. Sometimes, for very good and valid reasons, considerable compromise is necessary on the part of the manufacturer in the creation of a USAN, but USAN staff will work diligently to make that compromise as palatable to the drug innovator as possible.

INTERNATIONAL NONPROPRIETARY NAMES

No discussion of nonproprietary name development in the United States is complete without recognizing the liaison relationship that exists with the INN Committee. All names coined as USANs are sent to the WHO for consideration by its Committee. The USAN Council, the manufacturer, and the INN Committee negotiate to establish USANs as INNs and INNs, where appropriate, as USANs.

Under its charter, WHO is empowered simply to recommended specific actions or procedures to its members states. WHO first publishes its selected names as "proposed international nonproprietary names" (pINNs). Member states or other interested parties are allowed four months to submit comments or objections to any proposal. If no objection is received, WHO changes the status of the name to a "Recommended Nonproprietary Name" (rINN). Both recommended and, depending upon circumstances, proposed names are used by many member states as the sole or preferred nonproprietary name. All INNs created since 1953 are listed in the *USAN and the USP Dictionary of Drug Names*. A cumulative list of INN and the guidelines for creating them can be obtained from WHO in Geneva, Switzerland.

SUMMARY

Nonproprietary names serve a useful function that cannot be supplanted by the chemical name, trademark, or manufacturer's codes. National and international committees work in concert to coin nonproprietary names in an orderly fashion and establish global names. It is obvious to these committees throughout the world that if there is to be a workable and logical system of name development for use in the public domain, all regulatory agencies and interested parties, including the pharmaceutical industry, must work together. Trademarks and nonproprietary names should each stand alone as independently developing systems. To let one intrude upon the other will introduce confusion into an orderly system of drug nomenclature that has taken years and an inestimable number of man hours to refine.

Nomenclature of Biotechnologically Derived Pharmaceutical Substances – A View from Japan

TAKAO HAYAKAWA

National Institute of Hygienic Sciences,
1-18-1 Kamiyoga, Setagaya-Ku, Tokyo 158

Abstract: Some aspects regarding Japanese Accepted Names (JAN) for biotechnologically-derived pharmaceutical substances are described from the point of view of a regulatory scientist who is engaged in the quality control of these types of drugs, as well as from the viewpoint of a biochemist.

PROCEDURES FOR REVIEW AND DETERMINATION OF JAN FOR NEW DRUGS

In Japan the regulatory agency for drug approval is the Pharmaceutical Affairs Bureau of the Ministry of Health and Welfare. However, the scientific and clinical aspects of the applications for new drugs for approval are essentially reviewed by the Central Pharmaceutical Affairs Council (CPAC), an advisory body to the Minister of Health and Welfare. The CPAC has many review boards which are composed of medical and pharmaceutical specialists who are respectively engaged in the examination of new drugs according to their category. A JAN for a newly developed pharmaceutical substance intended for human use in Japan is reviewed for its application and determined by the Subcommittee on Japanese Accepted Names. Since the Japanese government is greatly interested in holding further discussions with other countries to coordinate positions with regard to the establishment of nonproprietary names for new drugs, the CPAC generated JAN will be applied to further international review through the International Nonproprietary Names (INN) committee of the World Health Organization (WHO). If the JAN is accepted by this committee, it is designated as a proposed INN, and will be published in the WHO Drug Information and also relayed to member countries. If there is no formal objection within 4 months after publication, the proposed INN will become a recommended INN. In some cases, the CPAC generated JAN may not be established as an international name. Even

K. Loening (Editor)
The Terminology of Biotechnology:
A Multidisciplinary Problem

in such cases, the Japanese CPAC committee will make every possible effort to reexamine the original JAN so that WHO recommended INN is adopted as the final established JAN. The Minister of Health and Welfare makes the final decision on designation in accordance with the CPAC report.

SOME ISSUES CONCERNING NOMENCLATURE OF BIOTECHNOLOGICALLY-DERIVED PHARMACEUTICAL SUBSTANCES

The development of biotechnologies like recombinant DNA technology, hybridoma technology and cell culture technology has enabled a wide variety of new medically useful agents to be produced in adequate quantities for preclinical and clinical studies or product marketing. To have biotechnology products contribute significantly to the public health, it is essential that suitable measures, based on the latest sound scientific principles, be taken by the manufacturers and control authorities to ensure the quality, safety and efficacy of these products. In relation to this, it has also become essential that some new principles for the determination of nonproprietary names for these biotechnology drugs be established.

It is said that names and nature often agree. In the case of pharmaceutical substances intended for human use, the names and nature should correspond. Under a given nonproprietary name, we can identify a common substance with respect to the chemical structure or specific chemical and biological features. This specific structural feature or common characteristics is a basic principle which will prescribe the quality, safety and efficacy of a given substance as a drug. Therefore, in principle, drugs which possess different chemical structures or characteristics should be distinguished from one another in terms of the individual nonproprietary name. The converse also holds true. Substances which possess the same chemical structure should be unified into a common name. Thus, the nonproprietary names of drugs play an essential role by which we can identify or distingish an individual drug which possesses peculiar characteristics, safety and efficacy for clinical use. Of course, as Dr. Schiffman has described in the preceding paper, nonproprietary names do serve a function that cannot be supplanted by the chemical name, trademark, or manufacturer's code, and they should be simple, short, unique, and useful names that may be used without restriction in the public domain.

When we take these points into account, in order to select suitable nonproprietary names for biotechnologically-derived pharmaceutical substances, what are the new issues which must be considered? What are the new principles to be established?

Specific Features Of Biotechnology Drugs

One of the most specific features of biotechnology drugs with respect to their identity which may become a problem for the selection of nonproprietary names will be found in the manufacturing process. The production of drugs by biotechnology is based on the artificial manipulation of a gene or/and a living organism. There may be variability, as well as options within the manufacturing process. There are many steps or factors involved in manufacturing processes for biotechnology drugs, which may affect the identity and characteristics of the final product. Due to the ease of changing nucleic acid codons, site-specific modification of proteins can often be done at will. A wide variety of scenarios with respect to the construction of recombinant DNA, host cell selection, the establishment of recombinant cells or cell lines as a seed lot for drug production, culture conditions, conversion of the primary product to the final product, and the purification process may also be left to the manufacturer's discretion. On the other hand, it is well known that genetic, posttranslational or any other processing will be controlled by many factors, such as the host cell-type or culture conditions described above.

This leads to the production of not only a natural form or site-specific mutant having the intended structure, but also any variant or derivatives. The latter include: a genetic variant, N-Met-derivative, glycotype variant, N-deacyl-derivative, C-desamido derivative, proteolytic cleavage product such as truncated forms at the N- or C-termini, or the clipped form (the internal peptide bond cleft two chain-form), a mixture of subtypes like interferon alfa, precursor form, and so on.

Nomenclature Of A Specific Simple Protein Drug Having A Specific Intended Structural Feature

Among these, in the case of a simple protein, either the natural form or its mutant, which has a specific intended structural feature including amino acid sequence, it is rational to give an individual nonproprietary name to it, based upon its specific biological properties or scientifically adopted name. In this case, a specific simple protein having the same amino acid sequence should be given a common nonproprietary name regardless of which manufacturing processes have been used, while the technology used, such as genetical recombination or cell culture, may be given in parentheses after the common nonproprietary name, as is ruled for JAN.

Nomenclature Of A Genetic Or Posttranslationally Modified Form

How shall we treat the case of a genetic or posttranslationally modified form, in particular, a simple protein that may have only a minor variation of a highly complex macromolecule produced by expression of a common structural gene? Is it rational to distinguish it from its natural form or originally intended form with respect to terminology? From the point of view of strict chemical structural features, the answer is yes. How about the criteria from the view point of safety and efficacy of such a product as a drug? The answer may also be yes. Some examples which may suggest the rationality of such treatments can be found in the following findings.

At first, as an example of a minor variation in the primary structure of a biotechnology product, it seems to be valuable to touch upon the effect of an additional methionine residue at the amino terminal of a polypeptide on safety or efficacy. It has been reported in a clinical trial with a recombinant methionyl human growth hormone, that the product was more antigenic than its pituitary counterpart hGH (1). The additional amino terminal methionine is an obvious possible cause. In relation to this, it is also noteworthy that various types of insulin which differ one another by one or three amino acids, respectively, have different antigenicities in clinical use. On the other hand, it has been found that the presence of a methionine residue in the amino terminal portion of IL-1 beta affects the biological activity. The *in vitro* biological activity of Met-IL-1 beta is only one-tenth of IL-1 beta or its desalanine derivative, as measured by the ability to inhibit the growth of human melanoma A375 cells (2). These results suggest that a change in even one amino acid residue in a protein may have the potency to modify the biological features of its counterpart product with respect to safety and efficacy.

An another intresting example is found in the relationship between human urokinase (UK) which is an active form of plasminogen activator and single-chain pro-urokinase (p-UK) which is an inactive proenzyme form. It was found that upon activation by a catalytic amount of plasmin, single-chain p-UK was cleaved at the Lys-Ile bond between residues 158 and 159, resulting in a two-chain UK molecule form held together by one disulfide linkage (3). Although the amino acid sequences of both proteins are essentially identical, this cleavage induces a conformational change, which is followed by the generation of plasminogen activator activity and the reduction of high affinity for fibrin of the single-chain form; the affinity of single-chain p-UK for fibrin clots is much higher than that of UK (3). In vitro thrombolytic studies show that single-chain p-UK is approximately three times more potent with respect to fibrinolysis than UK and that it does not degrade fibrinogen in the plasma at a concentration, at which complete plasma clot lysis takes place. On the other hand, UK extensively degrades fibrinogen in the plasma (4). These specific potent thrombolytic properties of single-chain p-UK seem to be due to its high affinity for fibrin and to its conversion from the inactive single-chain form to the active two-chain form on the thrombus by the catalytic amount of plasmin generated during coagulation. This single-chain p-UK might be more advantageous than UK in thrombolytic therapy. Thus, these two types of plasminogen activators should be distinguished from each other in terms of nonproprietary names, even though the amino acid sequences of both proteins are essentially identical.

Aspects Of Biotechnologically-Derived Glycoprotein Drugs

Many medically useful glycoproteins have been produced by biotechnology and some are already under application for new drug approval. One of the points to be considered when evaluating glycoproteins produced by biotechnology as pharmaceutical substances is that oligosaccharides attached to a protein may up-regulate, down-regulate, or turn off the biological activity of the glycoprotein in question. They may also affect the metabolic fate, stability or solubility of the protein. There are many examples of the biological activities of glycoproteins being controlled by the nature of glycosylation, as well as by the occupancy of glycosylation sequons (5-6). The presence and appropriate structures of N-linked oligosaccharides of (probably alfa-subunit) human gonadotropic hormones such as hCG, LH, FSH, and thyroid-stimulating hormone are essential for the activation of adenyl cyclase in the target tissue - but not for binding to the hormone receptor (7-11). Both the occupancy and nature of the oligosaccharide at the N-glycosylation site of tissue plasminogen activator (t-PA) can influence the rate at which they form an active complex with fibrin, a complex which is able to cleave plasminogen (12-13). It is well known that the nature and extent of glycosylation of erythropoietin (EPO) may affect its biological activity in vivo (14-17).

Further points to take into account are that species-, tissue- and cell-specific N-glycosylations are now firmly established, and that the unique glycosylation capacity of the cell has been referred to as defining its glycotype (5, 13, 16, 18-21). A typical example of cell-type-specific and site-specific N-glycosylations has recently demonstrated in the case of Type I and Type II t-PA from a human colon fibroblast cell strain and a Bowes melanoma cell line. The t-PA glycoprotein is secreted by each cell line as a set of glycoforms, each glycoform being unique with respect to the nature and disposition of oligosaccharides on a common polypeptide. Two cell lines express no common glycoform, despite expressing the same t-PA polypeptide (18). Similarly, it has been revealed that Chinese hamster ovary (CHO) cells derived recombinant t-PA is N-glycosylated differently from murine C127 derived recombinant t-PA (13). Both qualitative and quantitative differences in t-PA N-glycosylation influence its enzymatic activity (12-13). Thus, the expression of glycoproteins of one species in another, by genetic engineering, may lead to the isolation of variants of that glycoprotein. Even within the same cell type, the production of distinct sets of glycoforms which differ with respect to the fine structure of certain oligosaccharides, and the relative incidence of common oligosaccarides may occur, as seen in CHO cell-derived EPOs (17, 22-23). Many factors such as the method of establishment of the cell seed, culture conditions used or structural features of expressed protein product can exert a certain influence on the nature and extent of glycosylation (19, 22-27).

Proteins without the appropriate carbohydrate moieties may have altered pharmacokinetics and tissue distribution (7, 14, 28-32) and also may cause other unknown and possibly adverse effects, including an antagonistic action against its natural form (8, 11, 33), immune complex formation through interaction with naturally occurring antibodies which recognize a specific carbohydrate structure (20), possible activation of monocytes (34), and interaction with various cellular receptors for its carbohydrate moiety (5). For those products which are administered parenterally in large amounts or repeatedly over long periods, nonglycosylated proteins may reveal new antigenic sites while glycoproteins with unnatural carbohydrate structures may be antigenic.

From the point of view of being chemically distinct as well as safety and efficacy, there is no doubt about the importance of distinction of glycoprotein and glycopeptide drugs produced by a different seed cell in terms of nonproprietary name, even if they possess a common polypeptide backbone.

SUMMARY

As we have learned so far, it seems to be rational to establish some principles which distinguish JAN for biotechnologically-derived pharmaceutical substances. These are as follows:

1. A simple protein or polypeptide drug which is shown to be entirely identical with respect to its specific structural features including its amino acid sequences and the arrangement of disulfide bond(s), if present, to that of a naturally occuring drug or a previously approved drug made by any other method should be given the identical JAN.

2. A simple protein or polypeptide drug which is produced by expression of a common structural gene, but is shown to be modified during genetic, posttranslational, and/or any other processing should be distinguished by JAN. However, in such a case, the prefix, suffix, or stem may be common.

3. A simple protein or polypeptide drug which is produced by expression of a different structural gene should be the subject of distinction by JAN. However, in a case such as interferon subtypes, they may have a common stem and be distinguished only by the hyphenated addition of a number and a letter, such as interferon alfa-2a or 2b.

4. Glycoprotein or glycopeptide drugs should be the subject of distinction by JAN. However, in the case of glycoproteins which possess the same polypeptide backbone but a different or an unknown carbohydrate structure, they may have a common name and

the distinction may be made only by the addition of a letter or the hyphenated addition of a number and a letter.

5. In all cases, the technology used, such as genetical recombination or cell culture, should be given in parentheses after the name.

6. In principle, the term of the stem may be an abbreviated form of the term of the naturally occurring counterpart which is well established in the scientific literature. In the case of biotechnologically-derived glycoproteins, the term itself for natural substances cannot be designated because of potent variations in glycosylation.

7. Wherever possible and appropriate, information on the type of drug, such as a simple protein, polypeptide, glycoprotein or glycopeptide, molecular weight, the number of amino acid residues, and the name and/or origin of the structual gene or established cell lines should be provided as a specific feature. In the case of a glycoprotein, it is desirable to provide the name of the seed cell.

It is our hope that appropriate nomenclature for biotechnologically-derived pharmaceutical substances will result in a certain mechanism by which we can identify or distinguish an individual drug which has peculiar characteristics, safety and efficacy for clinical use, and communicate and trade at the international level with a common understanding. These nonproprietary names also serve the function as a base upon which appropriate reference standards for a given biotechnology drug is established.

References:

1 Kaplan SL et al; Bierich JR (1986) In: Raiti S, Tolman RA (eds) Growth hormone. Plenum Medical, New York London
2 Kikumoto Y et al (1987) Biochem Biophys Res Commun 147:315
3 Kasai S et al (1985) J Biol Chem 260:12377; Kasai S et al (1985) J. Biol Chem 260:12382
4 Kasai S et al (1985) Cell Struc Func 10:151
5 Rademacher TW et al (1988) Ann Rev Biochem 57:785
6 Knight P (1989) Bio/Technology 7:35
7 Moyle WR et al (1975) J Biol Chem 250:9163; Goverman JM et al (1982) J Biol Chem 257:15059
8 Nishimura R et al (1981) J Endocrinol Invest 4:349; Mizuochi T et al (1983) J Biol Chem 258:14126
9 Sairam MR, Bhargavi GN (1985) Science 229:65
10 Berman MI et al (1985) Biochem Biophys Res Commun 133:680
11 Matzuk MM et al (1989) J Biol Chem 264:2409
12 Wittwer AJ et al (1989) Biochemistry 28:7662
13 Parekh RB et al (1989) Biochemistry 28:7670
14 Goldwasser E et al (1974) J Biol Chem 249:4202
15 Ochi N et al (1987) Exp Hematology 15:556
16 Goto M et al (1988) Bio/Technology 6:67
17 Takeuchi M et al (1989) Proc Natl Acad Sci USA 86:7819

18 Parekh RB et al (1989) Biochemistry 28:7644
19 Baenziger JU, Green ED (1988) Biochim Biophys Acta 947:287
20 Kagawa Y et al (1988) J Biol Chem 263:17508
21 Tsuda E et al (1988) Biochemistry 27:5646
22 Sasaki H et al (1987) J Biol Chem 262:12059
23 Takeuchi M et al (1988) J Biol Chem 263:3657
24 Ohkura T et al (1984) Biochemistry 24:503
25 Opdenakker G et al (1988) FEBS Letter 238:129
26 Rothman RJ et al (1989) Biochemistry 28:1377
27 Lustbader J et al (1987) J Biol Chem 262:14204
28 Morell AG et al (1971) J Biol Chem 246:1461
29 Donahue RE et al (1986) Cold Spring Harbor Symposia Quant Biol 51:685
30 Lucore CL et al (1988) Circulation 77:906
31 Collen D et al (1988) Blood 71:216
32 Kinoshita H et al (1989) In: Marshak DR, Liu DT (eds) Current Commun Mol Biol (Formulation, Delivery, Targeting). Cold Spring Harbor Lab, New York
33 Dahl KD et al (1988) Science 239:72
34 Suzuki K et al (1985) Chem Pharm Bull (Tokyo) 33:886

Terminological Activities in Immunology

T. Tada and S. Matsuoka

Department of Immunology, Faculty of Medicine, University of Tokyo,
Hongo 7-3-1, Bunkyo-ku, Tokyo 113, Japan

Abstract: The immune system is composed of a large variety of cells and molecules which are characterized by their diverse biologic functions. In order to describe their generation and the dynamic interactions, meaningful and coherent terminology is absolutely necessary. However, the situation is yet unsatisfactory because of the reasons we describe here. The mother society of immunological organizations, the International Union of Immunological Societies (IUIS), has started a nomenclature committee in 1973, and has organized a series of committee meetings where a standard procedure for immunological terminology has been proposed. The procedure consists of 5 steps: 1) Establishment of a specific committee (working group), 2) development of an initial draft, 3) review by parental WHO-IUIS nomenclature committee, 4) submission of the document to the WHO Bulletin, 5) submission to other journals using the reference from the WHO Bulletin. Several activities are currently on-going (1). However, the problems have become more complicated than the time of above proposal as we describe here, which do not allow the simple application of the above procedure. We shall discuss in this communication how the nomenclature problems have arisen in immunology, and what efforts are being made to solve them. The problems in immunological terminology are probably typical ones which are shared by other disciplines of molecular and cellular biology. Three important areas were chosen to present current movements to establish unified nomenclature in immunology. They are 1) HLA (human histocompatibility antigens) by the International Histocompatibility Workshop and Conference, 2) CD (leukocyte differentiation markers) by the International Workshop and Conference on Human Leukocyte Differentiation Antigens, and 3) IL (interleukins-lymphokines) by the International Lymphokine Workshop and IUIS nomenclature subcommittee.

HLA-HUMAN HISTOCOMPATIBILITY ANTIGENS AND THEIR GENES

HLA was initially a designation for human leukocyte antigens detected by allo-antibodies mainly derived from patients receiving repeated blood transfusion or from pregnant women who developed antibodies against non-erythrocyte antigens

K. Loening (Editor)
The Terminology of Biotechnology:
A Multidisciplinary Problem

or fetal tissue antigens. It was at first believed to be a leukocyte alloantigen, as the antibodies were detected by agglutination or lysis of peripheral blood leukocytes. Nothing was known about the locus and allelic form of genes for the antigen. Later on, HLA antigens were found to be multiple, and some are expressed on virtually all somatic cells, not restricted to leukocytes. There are two types of HLA antigens, now designated as class I and class II. Class I antigen is composed of heavy and light chains, and the polymorphism defining the histocompatibility in transplantation lies on the heavy chain. Light chain is a monomorphic β_2 microglobulin. There are 3 loci which encode class I heavy chain, HLA-A, B and C, each of which has a number of alleles that define the polymorphism. The designation of HLA class I antigen has mostly been based on the serological specificity of antibodies, which generally refects, but not always faithfully predicts the differences in the amino acid sequence of heavy chain. Now, the polymorphism in heavy chain genes can be precisely determined either by the restriction fragment length polymorphism (RFLP) or by the use of DNA probes.

Class II antigens are composed of α and β chains, combination of which produces serological polymorphism. We know that there are three well defined chromosomal regions, HLA-DP, -DQ and -DR regions, where several α and β chain loci are mapped. Three other loci have been determined at the DNA level but not by the protein. Class II antigens are expressed only on certain tissues, macrophages, B lymphocytes and some epithelial cells. Class II antigens can be detected by serological tests as well as by the methods called mixed lymphocyte reaction (MLR), in which T lymphocytes undergo proliferation upon recognition of foreign class II antigens. They are sometimes called "lymphocyte-defined (LD)" antigens. Again, the designation of class II antigen was dependent on their reactivity with a panel of human antisera and lymphocytes from particular individuals. A large number of HLA class II types have been determined by such serological and cellular examinations.

This situation, where various immunological tests were used to define HLA types, unavoidably demanded an international consensus for the designation and comparative examination of data with standard antisera or lymphocytes. Those scientist working in HLA have organized meetings and workshop every three years to establish an internationally acceptable designation of HLA antigens and genes. The meeting is called "International Histocompatibility Workshop and Conference", which discusses the feasibility of nominating new antigen encoded by HLA loci.

Report of these meetings have been published in WHO Bulletins, with which where new nomenclature was finally accepted (reviewed in 2).

The nomenclature problem of HLA is a typical one in biological science. At first the designation of HLA was based on the serological specificity of leukocyte antigen detected by human standard antisera, most of which were derived from pregnant women. Then, some cellular techniques were introduced to define the LD antigens which mainly detect the HLA-DR (Class II) polymorphism. Monoclonal antibodies have also been introduced to facilitate biochemical characterization of class I heavy chain and class II α and β chains especially by two dimensional gel analyses. The introduction of molecular genetic techniques enabled the scientists to provide the definitive definition of HLA alleles. Southern blot analysis and restriction fragment length polymorphism (RFLP) with HLA-DNA probes have been used to define HLA class I and class II region genes more precisely. The nucleotide sequences of HLA genes and pseudo genes have finally defined the gene organization and allelic forms of HLA complex.

Table 1 depicts the flow of the HLA nomenclature which may be a typical example in biological sciences. It started with serological definition of antigens followed by biochemical designation of their components. It is now designated by molecular terms of loci (alphabet) and alleles (four digit number) including pseudo genes which could never been detected by antibodies.

Table 1. The flow of nomenclature in HLA system

Methods of detection	Designation	
Serological	HLA antigen	
	HLA-A, B, C, (DR)	
(+ cellular)	HLA-A, B, C	D (LD)
Biochemical	HLA-A, B, C	DP, DQ, DR
(+ monoclonal antibody)	(class I, α chain)	(class II, α and β)
Molecular (regions)	A, B, C, E(ψ),	DP, DQ, DR, DN, DO
(loci)		A(α) B(β)
DNA sequence (alleles)	A*0000	A*0000 B*0000

CELL SURFACE MARKERS AND CD NOMENCLATURE

One of the major achievements in modern immunology was to establish that the immune system is composed of a variety of lymphoid cells having diverse functional properties. They are, however, derived from a single cell type, stem cell in the bone marrow, which differentiates into a wide variety of functional cell types including peripheral blood cells, macrophages, T and B lymphocytes. T lymphocytes are divided into subpopulations having different functions, such as cytotoxic, helper and suppressor T cells. B cells and macrophages are also known to undergo sophisticated differentiation. During the differentiation processes they express various characteristic glycoproteins on the cell surface, which were initially called cell surface markers. The cells in the immune system are making mutual interactions to regulate the function, differentiation, growth and maturation of other cell types. In order to describe the generation and dynamic interaction of these cells, it is necessary to designate the cell types by their characteristic cell surface markers.

When immunologists first realized the existence of functionally different cell types in the immune system in 1960th, at first T and B cells, their desire was to establish a method to physically distinguish these cell types. They were successful to obtains alloantibodies to distinguish T and B cells at first, and then those which can distinguish functional subpopulation of T lymphocytes, helper and killer lymphocytes. They thought that epitopes detected by antibodies are mere surface markers and have nothing to do with the function born by these cell types. In general, such marker molecules detected by alloantibodies were designated by the putative alleles determining them, such as Ly 1.1, 1.2, and Thy 1.1, 1.2, etc.

Later on a large number of monoclonal antibodies were produced which reacted with cell surface molecules of certain lymphoid cells but not others. Several of these molecules were found to bear important cellular functions in cell-cell adhesion and in signal transduction. By the combination of monoclonal antibodies, immunologists have been able to define the pathways and steps of differentiation of lymphoid cells, and to distinguish each small subset having subtle functional differences. The rapid development of microfluorometry in conjunction with the commercial availability of monoclonal antibodies enabled us to classify diverse subpopulations of lymphoid cells in the whole immune system. This was undoubtedly a great advantage in studying this highly complex biologic system.

However, confusion has started with this new achievement. Many commercial farms started to sell their antibodies reacting with certain lymphocyte subpopulations, which are coded by their own commercial names. Different firms gave different arbitrary names which appeared in a large number of scientific papers. Problems have arisen, as the same differentiation molecule could produce different antibodies which recognize different epitopes on this same single molecule. For example, an antibody from a company A reacts with the epitope A and the other antibody from company B recognizes the epitope B on a same single molecule on an identical cell type. One author defines the cell types by the A epitope, and other by the B epitope. If there are no critical comparative efforts, the whole area should become caotic.

To solve this situation, scientists in this field has started a workshop "International Workshop and Conference on Human Leukocyte Differentiation Antigens" in 1982, where the monoclonal antibodies used to define differentiation antigens were critically compared, and those detecting the same molecule were classified as a "cluster of differentiation (CD)". This awkward name was at first to designate a group of monoclonal antibodies which detect the same differentiation-associated molecule, but nowadays used to designate antigenic molecule itself. This is a natural consequence of initial desire of immunologists to designate the differentiation antigen rather than the antibodies. It is, therefore, often used as "CD antigen".

Since 1982 four conferences have been organized, and successfully renamed many cell surface differentiation markers by the CD nomenclature (3). At the recent 4th International Workshop in Vienna (1989), 78 CD numbers have been assigned to surface antigens of human leukocytes (4).

The process to assign a CD number is relatively simple; the monoclonal antibody and its data are at first sent to the head office of the workshop. They are then distributed to multiple registered laboratories and examined on defined criteria in blind. The results are returned to the head office, and data are analysed by a computer to assign new CD number. These data are discussed at the workshop for their feasibility to obtain the official designation, and finally the recommendation is published in widely circulating journals.

This new terminological activity gave a tremendous advantage to define cells and cell surface functional molecules with the same international designation. All the scientists can speak the same language to discuss complex phenomena manifested by diverse cells. In addition, they can define CD antigens of other animal species by the molecular homology.

However, there are several problems with CD nomenclature. Since CD was first given to a group of monoclonal antibodies, not to the molecule itself, there is always a confusion in its usage. We often hear CD antigen and CD molecule, and sometimes CD antibody and anti-CD antibody etc. Secondly, CD is only a code and reflects no functional significance of the molecule. It can be used only within the guild or the circle of the same profession, but not to other scientists who are working in other specialities even in immunology. In many cases, immunologists use commercial and trivial names for their practical use, such as "T3 antigen" and "Fc receptor", and translate them into CD numbers when it is absolutely necessary.

It also suffers from the problem caused by the differences between serological and molecular characterizations. Many of the serologically defined molecules have now been biochemically identified, and their genes cloned. As the results, it was frequently found that serological determinants were sometimes shared by different molecules. The same single molecule was divided into two different isoforms. Some of the serologically defined antigens have been found to be heteropolymer composed of different polypeptides each of which has different epitopes. In addition, allo- and xeno-antibodies were found to define different degrees of heterogeneity. Some CD antigens had to be divided into two or more, such as CD1a and CD1b. These problems have not been completely resolved. The transfection of genes into host cells and recharacterization of antibodies with the transfected cells may be necessary in the future for absolute definition.

INTERLEUKINS AND LYMPHOKINES

Interactions between diverse cell types in the immune network are mediated by a series of soluble molecules generally called lymphokines or interleukins. Lymphokines are biologically active factors produced by lymphocytes. There are several functional molecules produced by non lymphoid cells such as macrophage, monocytes and neutrophiles. They are included in "cytokine family". The term "Interleukin (IL)" was also proposed to designate functionally active molecules mediating interactions between leukocytes. These names were first reasonably given for structurally unknown carriers of biological functions, but now cast several problems, which no more allow the innocent usage of cytokines and interleukins.

When the molecular characterization of these factors advanced, some of the functionally determined factors were found to be identical to each other. One single lymphokine was found to have many different functions. For example, a group of monocyte-derived factors acting on thymocytes and B cells to stimulate them, i.e., T lymphocyte activating factor (TAF) and B cell activating factor (BAF) were identical on the basis of biochemical similarities. They were renamed as IL-1. In addition, they found that the monocyte derived mediator that induce fever, namely endogenous pyrogen (EP), was the same molecule as IL-1. IL-1 was found to have several other properties; to stimulate liver cells to produce the battery of acute phase proteins and to induce migration of leukocytes from bone marrow to circulation. Thus, IL-1 has multiple functional activities acting on multiple target cells. Such redundancy and ambiguity of the same molecule do not allow the use of functional names. Another unrelated lymphokine, later designated as IL-6, had a few similar biologic properties to IL-1 in stimulating liver cells and inducing fever, etc. Such a degeneracy in their functions was not compatible with the existing functional names.

Let us look at a symbolic example of the process of naming IL-6. In 1983 after the 5th International Congress of Immunology in Kyoto, several immunologists studying the T lymphocyte-derived molecules which stimulate B lymphocytes to proliferate and differentiated into antibody forming cells. Such molecules had been called as T cell replacing factor (TRF), B cell growth factor (BCGF), B cell differentiation factor (BCDF) etc. It was proposed at this meeting, that factors which had been well characterized functionally and chemically, preferably by more than one laboratory, should be given a formal designation. B cell stimulatory factors (BSF) were the proposed names, and they should be sequentially numbered such as BSF1, BSF2 etc. If the factor is not purified to homogeneity or its structure is not determined from cloning and sequencing of its gene, small p indicating provisional should precede the number (5).

Thus, the name of BSF-2 (or BSF-p2) was given to one of the functional molecule, BCDF. Five years later, however, those scientists re-assembled at a New York Academy of Sciences meeting, and renamed BSF-2 as interleukin 6 (IL-6). During this time period, structural analyses of this lymphokine and other biologically active substances have been made, and as a consequence, BSF-2 was found to be structurally identical to a few other functionally defined lymphokines, such as, hybridoma-plasmacytoma growth factor (HPGF), hepatocyte stimulating factor (HSF), interferon β_2 (IFN-β_2), 26 kDa protein, TRF, etc. These molecules are produced by T cell, B cell, macrophage, fibroblast, keratinocyte, endothelial cells, and even by glial cells and astrocytes, and act on a variety of immune and non-immune cells. Such redundancy and degeneracy caused a great difficulty of the nomenclature for these polyfunctional molecules. Because of the diverse sources and diverse targets of the molecule, the term "interleukin" was not very adequate. However, those scientists finally accepted "IL-6" abandaning their own functional names. After this historical event, three other interleukins, IL-7 to IL-9, have been adopted, most of which are also polyfunctional molecules not fitting in the initial concept of "Inter-leukin". The use of interleukin designation for molecules that have many actions and many target cells other than leukocytes should not be satisfactory in the future, as we expect the existence of many regulatory molecules commonly used in immune, endocrine and nerve systems.

Is there any better solution for this? The nomenclature committee and its lymphokine subcommittee of IUIS have been making efforts to settle the problem. The proposal given by a member of the committee, Dr. William E. Paul of National Institute for Allergy and Infectious Diseases, Bethesda, MD, is worthwhile to mention.

On the basis of a series of discussion with large number of scientists he proposed that IL nomenclature should have a great advantage of wide acceptance and wide popularity, and could be used as a new code system to designate polyfunctional molecules. He proposes that new factors should be assigned with new IL numbers, once they have been purified, cloned and characterized. The existing factors should also receive official IL descriptors, although the trivial names could be in principle changed. Thus, interleukin 2 would be known as interleukin 2 (IL-2), where IL-2 is an official descriptor and interleukin 2 would be the official name. Other names for the molecule could be used as long as the official descriptor is appended. Thus, new molecules, as well as those known to be a member of IL families, should be given official descriptors. For example, granulocyte-monocyte colony stimulating factor (GM-CSF) would be IL-n (but not interleukin-n) and IFN-γ would be IL-n+1. Such a designation should be given by the IUIS nomenclature committee but should not be adopted unilaterally. The working group of IUIS subcommittee for an IL descriptor should be in function to make official pronouncement.

CONCLUSION

We have discussed the terminological activities in immunology and the background of problems with reference to three representative areas. We see here is an unavoidable fate of terminology in biological sciences. As shown in Table 2, the first terminology came from functional properties which was iconic to the reality of the molecule, but was always redundant. The second step is the serological and structural designations. It will finally adapt the code system based on the molecular information. The signs and symbols will be replaced by the codes. The end is, alas, undoubtedly an abstract jargon being so apart from biologic significances.

Table 2. Fate of terminology in immunology

Principle	Mode	Problem
Functional	Icon	Redundant
↓	↓	Degenerative
Serological	Sign	Ambiguous
Polyclonal		Arbitrary
↓	↓	
Monoclonal		Incoherent
Structural	Symbol	↓
↓		Incomprehensive
Molecular	↓	Unrealistic
↓		↓
	Code	Artificial
		Abstract
		Jargon

ACKNOWLEDGEMENTS: We thank Dr. B. Cinader, the chairman of the IUIS Nomenclature Committee, Dr. W. E. Paul, NIAID, NIH, Bethesda, MD, Dr. T. Sasazuki, Kyushu University, Fukuoka, and Dr. M. Miyakasa, Tokyo Metropolitan Institute for Medical Sciences, Tokyo, for thier information about activities in different areas.

REFERENCES

1 Cinader B (1984) Int Arch Allergy Appl Immunol 95:193
2 Bodmer WF, et al (1989) In: Bodmer B (ed) Immunobiology of HLA. vol 1, p 72, Springer Verlag, New York
3 McMichael AJ et al (eds) (1987) Leukocyte Typing III. Oxford University Press, Oxford
4 Haynes BF, Denning SM, Siger KH, Kurtzberg J (1989) Immunology Today 10:87
5 Paul WE (1984) Immunogenetics 19:33

Trends in Standardization of Vocabulary in Microbiology

Micah I. Krichevsky

Microbial Systematics Section, Epidemiology and Oral Disease Prevention Program, National Institute of Dental Research, National Institutes of Health, Bethesda, MD 20892

INTRODUCTION

Historically, biologists have established formal rules for the naming of organisms. The drive to standardize the naming conventions in biology, including microbiology, began long before the advent of computers and the attendant construction of databases. The wish to classify gave the impetus to the creation of procedures and standards for publishing and legitimizing the names of taxa.

Many glossaries of terms used to describe attributes of organisms have been published over the years. However, the efforts in this area have been sporadic in both time and taxa domains. Very few legitimizing conventions exist. The International Committee on the Taxonomy of Viruses (ICTV) has published a compendium of descriptive terms[1]. While "official", this compendium has seen but limited use.

Informal adoption of standard terms from other disciplines occurs. Phycologists and mycologists often use the names of plane shapes (primarily for description of leaf shapes) codified by the Systematics Association of Great Britain[2,3,4]. The Nomenclature Committee of the International Union of Biochemistry provides standard names for many enzymes[5].

This discussion focuses on the above two areas. The existing procedures for naming organisms in the various branches of microbiology are illustrated by those used for bacterial names. Descriptive terms for attributes of microorganisms are discussed with respect to the informal authoritative compendia that exist as well as some thoughts on a specific effort at developing a "pan-microbiology" standard set of descriptors.

K. Loening (Editor)
The Terminology of Biotechnology:
A Multidisciplinary Problem

NAMES OF ORGANISMS

Names of organisms are assigned by the authors of the first paper describing the organism (or taxon). The description states the reasons why a new name should be created. The two most common reasons are discovery and characterization of a new organism or revision of the existing taxonomy.

The various branches of biology have rules for construction and validation of names. The botanists (including mycologists and phycologists) have a single Botanical Code of Nomenclature. Zoology has The Zoological Code of Nomenclature that includes protozoa. The International Code of Nomenclature of Bacteria covers these organisms. At one time the bacterial code included the viruses. The International Committee on the Taxonomy of Viruses has evolved a code specific to viruses[6].

These Codes of Nomenclature are not consistent in either rules or coverage. Some amoebae (e.g., cellular slime molds) and the photoflagellates (e.g., *Euglena*) are covered (coveted?) by both botanists and zoologists. John Corliss of the University of Maryland proposes establishing a single, new code of nomenclature for "protista" (this term includes the protozoa and algae, including the slime molds).

The words used as names of taxa must conform to the rules of Latin or Greek roots and declensions for most of the codes. Binomial names are commonly required (with some additions for special cases). The virologists nomenclature code has important differences from the others. Latin or Greek binomials are not required and even frowned upon in many areas of virology. Naming viruses after persons is not permitted.

FUNCTIONS OF CODES OF NOMENCLATURE

The various codes of nomenclature have most functions in common. All codes set forth the rules for assigning valid names to taxa. These include rules for etymology and precedence. None define the bounds of taxa. The concept of defining a species on the ability to produce progeny from mating clearly does not help with respect to microbes and also fails when rigidly applied to macrobiota.

The codes of nomenclature are strictly limited in application to the validity of the words used to name a group of organisms as a taxon.

They do not speak to the scientific validity of the application of the names. The codes allow conflicting taxonomies to exist. That adjudication of name validity also allows adjudication of taxonomic "correctness" is a common misconception of the aims and authority of the codes. The final decisions as to the "best" taxonomic schema rest with usage by the scientific community.

The code for viruses, as implemented by the ICTV, does differ somewhat from the others. The ICTV has reserved to itself more judicial power than those responsible for the other codes of nomenclature. The ICTV can and does change names of viruses to names other than those proposed by the discoverer of the virus. The naming of the Human Immunodeficiency Virus (HIV) is an example. Further, they will make scientific judgments on the validity of application of a name.

AUTHORITY FOR THE CODES

Most branches of science are represented in a hierarchical non-governmental organization, the International Council of Scientific Unions (ICSU). The constituent organizations of ICSU are the various international scientific Unions and various special interdisciplinary Committees and Federations. The authority for the codes of nomenclature derives from the authority of two Unions. The International Union of Biological Societies (IUBS) controls the zoological and botanical codes. The bacterial and viral codes are the domains of components of the International Union of Microbiological Societies (IUMS).

The component of IUMS responsible for the virus code is the aforementioned ICTV. The infrastructure for the bacteria is more complex with distributed responsibilities.

The IUMS created the International Committee on Systematic Bacteriology (ICSB). The ICSB revises and publishes the code itself. A number of advisory Subcommittees devoted to consideration of the taxonomy of specific taxa of bacteria report to the ICSB. The ICSB's International Journal of Systematic Bacteriology (IJSB) is the mandatory journal for notices of new or revised names of bacteria. While a description of a new taxon may be published elsewhere, at least a notice of the publication must be entered in the IJSB before the name is officially recognized.

The ICSB's International Judicial Commission adjudicates requests for nomenclature changes by interpreting the rules of the Code. The adjudication process is initiated by proposal and request for decision published in the IJSB.

Under the auspices of the ICSB, the "Approved Lists of Bacterial Names" was published in 1980[7]. Bacterial names not on the Lists were discarded as not official. Since then, all emendations of the Approved Lists have been published in the IJSB. Thus, no journal other than IJSB need be consulted for officially recognized names of bacteria.

"UNOFFICIAL" NOMENCLATURE STANDARDIZATION EFFORTS

The various bodies of ICSU sanction the Codes of Nomenclature and their ancillary activities. Many important, but unsanctioned, initiatives contribute to our ability to access and utilize existing nomenclature. These initiatives are compendia of nomenclature or taxonomically-based organism descriptions. Some examples will illustrate the richness of this body of information.

Zoologic Record

The Zoologic Record presents synopses of new taxa. While unofficial, it is a very good compilation of the growing nomenclature in zoology. There is no rule in zoology comparable to the notification in IJSB requirement of the Bacteriological Code. The Zoologic Record is almost complete and renders a like service.

Taxonomic Reference File

Biosciences Information Service (BIOSIS) maintains an online "Taxonomic Reference File" (TRF) of the bacteria gleaned from their extensive literature searching. The file contains both valid and discarded names of bacteria as well as the taxonomic relationships among the listed taxa. The file is managed and searched as a relational database. In addition to the online version, BIOSIS sells a version on floppy disk for use in personal computers.

World Data Center for Microorganisms

The World Federation for Culture Collections (WFCC) sponsors the World Data Center on Microorganisms (WDC) at the Institute of Physical and Chemical Research (RIKEN), Saitama, Japan. The WDC is an information center which supports culture collections and their users. The primary tasks are publication of descriptions of culture collections and the production of species-oriented directories of culture collection holdings. The first effort was the publication of a general World Directory of Culture Collections. Specific new directories include a World Directory of Algae, a World Directory of Protozoa, and a Directory of Asian culture collections.

The WDC currently holds information on almost 400 culture collections distributed over 56 countries. The core data of the WDC are descriptions of culture collections and their holdings allowing users to locate a culture collection and/or taxonomic category of microorganisms. They can find strains of specific species by consulting the list of species preserved in culture collections. The WDC database currently includes bacteria, fungi, yeasts, algae, protozoa, lichens, animal and plant cells, and viruses.

For information dissemination, the WDC uses a variety of communication media such as online searching (directly or through the gateway to the Microbial Strain Data Network), mail, telex, cable, facsimile, electronic mail, publication and magnetic devices (floppy disks and magnetic tapes).

The directories and databases of the WDC form another set of taxonomic reference files as well as a mechanism for location of type strains and type species on a worldwide basis. The culture collections are a necessary underpinning for applied nomenclature as the cultures provide characterized biological reference material for taxon validation.

Codes for Names of Genera of Bacteria and Yeasts

A systematic set of unique four letter codes for the names of genera of bacteria and yeasts have been developed as a combined effort of personnel of the Microbial Systematics Section, National Institute of Dental Research, NIH; BIOSIS and the American Type Culture Collection [8,9].

The purpose of these codes is to facilitate computer searching using a standardized, fixed length format as well as to provide a set of unambiguous codes to save space in publications. The codes were designed to be mnemonic. Largely, this last aim succeeded. However, current attempts to extend this logic to the fungi in general seems unlikely to keep the mnemonic character as there are over 1200 named genera of fungi. Perhaps a combination of letters and numbers will retain uniqueness even if the mnemonic character is lost.

ATTRIBUTES OF ORGANISMS

In contrast to the formalism of the rules and infrastructures for the nomenclature of microorganisms, almost no ICSU-based standard setting organizations exist to develop terminology for describing the attributes of microorganisms. The ICTV has published a code for the description of virus characters[1]. All other efforts at standardizing terminology are private efforts.

These efforts come in two main forms. The most common is a compendium of descriptions of large groups of taxa such as the bacteria, yeasts, algae, fungi, and protozoa. Such compendia often are internally consistent in their use of terminology and can be used as an effective source of standardized terminology. Some of these taxonomic compendia contain very useful glossaries (for example, that of Corliss[10]). The rarer form of standardized terminology is a specific dictionary for the terminology used in describing groups of taxa, such as the fungi[11].

Ancillary efforts at standardizing terminology exist that are useful in describing microorganisms such as for enzyme nomenclature and specific gene nomenclature. These vocabulary structures are only randomly utilized in the various glossaries and dictionaries.

Nevertheless, the glossaries and dictionaries are useful within their respective domains. Modern bio- and computer technology motivate the creation of vocabularies that are applicable across broad interdisciplinary domains. An integrated vocabulary for microbiology should find wide application.

WORDS

Words may be used in various ways across disciplines. A definition in one group of taxa either does not apply to other taxa or has different meanings. "Spore" may refer to a reproductive or a survival structure or both in the various microscopic life forms.

Synonymy causes even more complications. The terms lunate, arcuate, bow-shaped, and crescent-shaped are used to describe the same cell shapes. The microbial physiologists may not be overly concerned with the details of morphology. However, in spite of a precise set of rules governing chemical structure nomenclature, we use various trivial names almost exclusively. L-fructose, fructose, levulose, grape sugar all denote a single compound and none are the standard name.

Individual words are rarely sufficient to define a data element. A list of monosaccharide names does not by itself define the ability of microorganisms to catabolize carbohydrate monomers. The name of the carbohydrate must be associated with other words to convey usable information. We would like to know if the L-fructose is utilized (in yeast physiology), yields acid (streptococci, lactobacilli), gas (lactobacilli), etc. We define features implicitly in terms of linked combinations of words.

STATEMENTS

We need full statement definitions (one or more clear sentences) to give unambiguous meaning to the descriptors of microbial characteristics, for example:

"L-fructose is utilized."

"Acid is produced from L-fructose."

"Gas is produced from L-fructose."

A statement imparts meaning through the proximity, order, and context of the words in the statement. The proximity and order are controlled by the conventions of the language. Context considerations make statement vocabulary construction difficult. The builder must anticipate the effect of context on the user of the vocabulary.

The following two sentences with identical grammatical structures illustrate how context and the ability of the reader to interpret the meaning of the writer control the effectiveness of the communication.

We treated the animals with disease.

We treated the animals with penicillin.

Substitution of "penicillin" for "disease" significantly changes the meaning of the sentences.

A good vocabulary design captures primary observations where possible. For example, the ability to grow in various specific concentrations of sodium chloride are recorded rather than derived information. This point is illustrated as follows:

% NaCl	0	1	2	3	5	10	Derived Pattern
Strain A	-	+	+	+	-	-	Requires Na ion
Strain B	+	-	-	-	-	-	Inhibited by Na
Strain C	-	-	-	+	+	+	Halophile
Strain D	+	+	+	+	+	+	Halotolerant

The derived patterns cannot be interpreted to yield the exact pattern of ability to grow in the various concentrations of sodium chloride. Thus, the ability to make judgments from the recorded statements is largely unidirectional.

THE RKC CODE

Very few statement-oriented vocabularies exist. The "ICTV Code for the Description of Virus Characters"[1] is one example. Another is the so-called RKC Code initiated some time ago by Morrison Rogosa, Rita Colwell and me[12]. Both the effort to develop such a coding convention and the group contributing to the effort has enlarged markedly over the ensuing years.

The RKC Code has grown to encompass, with varying degrees of coverage, most cellular microorganisms: bacteria, fungi, protozoa, and algae. The main thrusts of the current activities are to expand the descriptors for fungi, include biotechnological (application) descriptors, allow for incorporation of genetic information into a strain record, formalize the coding of plasmid data, and begin to interface the system with virus descriptions.

The RKC Code is a statement oriented vocabulary. Each statement is a description of a possible strain attribute. Each statement is assigned an unique number. Some examples are:

C02001: What is your laboratory strain number or designation?
C02004: What is the species name (genus and specific epithet)?
003001: Cells are spherical (all perpendicular axes have ratios between 6:5 and 1:1).
003008: Cells are rod-shaped (cylindrical).
012021: Cells are gram positive.

012022: Cells are gram negative.
535115: The minimum inhibitory concentration (MIC) of penicillin G is ________ units/ml.
535188: Inhibitory zone diameter for penicillin G disk concentration 10 units is ______ mm.

The data encoded under the above categories is one of three types: character (alphanumeric) data, denoted by a "C" before the five digit statement number; numeric, encompassing statement numbers 500,000 through 999,999; and "binary" with statement numbers from 000,001 through 499,999. The character statements are further differentiated from the others by usually being couched as questions rather than declarative sentences.

A considerable portion of data recorded by microbiologists is in the form of presence/absence (yes/no). Any coding system must allow for missing observations. Thus, the data type is not truly binary but ternary in nature. Two states are the complement of a positive response in microbiology. That is, the complement of positive is the combination of negative and missing responses, i.e., NOT POSITIVE = NEGATIVE AND MISSING. All three responses are allowed as responses by the conventions of the RKC Code.

The RKC code is open-ended. The list of features (now numbering over 12,000) is continually growing with the contribution of many microbiologists throughout the world. The Editor of the RKC Code is Dr. C. McManus in our Section. Inquiries and suggestions on the use or expansion of the RKC Code are welcome.

Two main implementations of the RKC Code are current. The first is a database management system called the Microbial Information System (MICRO-IS). The MICRO-IS is a large suite of programs for entry, editing, managing, and analyzing microbiological data. The first version of the programs is installed on the computers of the National Institutes of Health and are used by a number of U.S. government agencies. An enhanced version of the programs is available for personal computers using the MS-DOS operating system. The latter programs are available from the Microbial Strain Data Network (MSDN), Institute of Biotechnology, Cambridge University, Cambridge, UK.

The second implementation is by the MSDN which provides a locator service for microbes and macrobiotic cell lines on a worldwide basis and facilitates communication of standardized data and other scientific information among microbiologists, cell biologists, and biotechnologists. The facilities of the MSDN are available for use by any interested scientist.

In the fields of microbiology, cell biology, and biotechnology there has been a lack of widely disseminated information on the existence of strains and cell lines with specific properties. What data were available were in isolated, non-comparable data sets. The MSDN fills this void by providing a method, through the building of the Central Directory, of rationalizing disparate data sets.

The Central Directory of data sources necessarily forces standardization since the RKC Code is used as the controlled vocabulary for the Directory. The Directory was the initial motivation for the creation of the MSDN.

Acceptance of the RKC Code, the MSDN Central Directory and the MICRO-IS as *de facto* standards depends on widespread use. The RKC Code is distributed through publication as a book[12] and supplementary articles in journals[13,14,15,16]. We are establishing a mechanism for distribution of the updated Code in ASCII text on floppy disks as well. Selected portions are accessible online to the public. Both the disks and online access are through the facilities of the MSDN.

REFERENCES

1 Atherton JG, Holmes IR, Jobbins EH (1983) Monographs in Virology 14:1
2 Systematics Association Committee for Descriptive Terminology (1960) Taxon 99(8):245
3 Systematics Association Committee for Descriptive Terminology (1962) Taxon 11:145
4 Systematics Association Committee for Descriptive Terminology (1962) Taxon 11:245
5 Nomenclature Committee of the International Union of Biochemistry (1984) Enzyme nomenclature. Academic Press, Orlando
6 Matthews RE (1982) Intervirology 17:1
7 Skerman VBD, McGowan V, Sneath PHA (eds) (1980) Approved lists of bacterial names. American Society for Microbiology, Washington, DC
8 Johnson R, Rogosa M, Krichevsky MI (1976) Int J Syst Bacteriol 26:278
9 Rogosa M, Walczak CA, Walat JM, McVey D, Krichevsky MI (1986) Int J Syst Bacteriol 36:464
10 Corliss JO (1979) The ciliated protozoa, 2nd ed. Pergamon Press, Oxford
11 Hawksworth DL, Sutton BC, Ainsworth GC (1983) Dictionary of the fungi. Commonwealth Mycological Institute, Kew, Surrey
12 Rogosa M, Krichevsky MI, Colwell RR (1986) Coding microbiological data for computers. Springer, New York
13 Daggett P-M, Krichevsky MI, Rogosa M, Corliss JO, Girolami JP (1980) J Protozool 27:353

14 Philpot CM, Krichevsky MI, Rogosa M (1982) Int J Syst Bacteriol 32:175
15 Jong S-C, Holloway L, McManus C, Krichevsky MI, Rogosa M (1988) Mycotaxon 31:207
16 Jong S-C, Ho HH, McManus C, Krichevsky MI (1989) Binary 1:187

Terminology in Pesticide Science

Izuru Yamamoto, Tokyo University of Agriculture,
Setagaya-ku, Tokyo 156, Japan

Terminological activities in agriculture are active in Japan. Each scientific society has authorized terminology pertinent to the science, taking the style of Japanese -English and English-Japanese dictionary. Pesticide Science Society of Japan also edited such a dictionary. Pesticides, implying " kill" has given a negative emotion to the public, while new pest control chemicals are often not bio-cides but regulating vital life processes and modifying pest behaviors to exert their action. An ACS division has changed its name to "Agrochemicals" from Pesticide Chemistry. The corresponding Japanese wording "NOYAKU" has already gotten a negative image. The growing number of chemicals which affect uncoventionally to pests and crops requires newer terminology such as bioregulators, bioregulants, pestistatics, behavior modifyer, behavioral chemicals etc. and provides a challenge to translate them into proper Japanese.

Terminological activities in agriculture are active in Japan. Each scientific society has authorized terminology pertinent to the science, taking the style of Japanese -English and English-Japanese dictionary. Pesticide Science Society of Japan has also edited such a dictionary.

To understand pesticides clearly, we first have to ask the question,' what is a pest?'. This is not easy. Three definitions are cited. 1) A plant or animal living where man does not want it to live(K. Mellanby, Pestiocides and pollution.) ; 2) A species that, because of its great numbers, behavior, or feeding habits, is able to inflict substantial harm on man or his valued resources(The Pesticide Conpiracy, R. van den Bosch); 3) an animal or plant that

K. Loening (Editor)
The Terminology of Biotechnology:
A Multidisciplinary Problem

occurs in such abundance as to present a distinct threat, economically or medically, to man or his interests(Encyclopaedia Britanica). It is obvious that an organism is not a pest until it does so at a level which is intolerable. What is intolerable to one person(community, country etc.) may be tolerable to another. One cockroch may be a pest for sensible lady.

The definition of Pesticides is also not easy. The word implied originally to kill pests. Before the World War II, chemicals had no remarkabe killing activity, and the capability of modern pesticides to kill pests were extraordinary and the term pesticides sounded fresh. However, for the past sevral decades pesticides meaning bio-cides has given a negative image to the public. Also, the growing number of chemicals which are not bio-cides but regulating vital life processes and modifying pest behaviors to exert their action requires newer terminology. An ACS division has recently changed its name to "Agrochemicals" from Pesticide Chemistry. The corresponding Japanese wording "NOYAKU" (agro-drugs) has been utilized since the beginning of modern pesticides after the World War II, but the public has already responded negatively to the wording. We need a better term, which reflects properly the so-called pesticides.

The definition of pesticides is changing and the terms included in a dictionary of pesticide science is ever increasing its number.·I would like to explain why it so happens.

In the past, the agriculture of old was carried out without the use of pesticides and despite all the efforts there were outbreaks of pests. Pesticides have provided: rapid curative action in preventing damage; flexible properties in usage; apparently simple and reliable control measures; and more freedom in other agricultural practices such as the use of fertilizers and felxible timing. These were the cause of the coming of pesticide era.

This means that terminology in pesticide science included all the terms related to the names and types of pesticides; physical states of pesticides; formulation, application techniques and machines, bioassay and responses of test organisms; all about adjuvants; all about analysis; pests and crops; regulations and legislation.

However, the assets of simplicity, reliability, efficiency, wide use and temporary economic return have themselves been the indirect cause of negative impacts, such as undesirable acute hazards to human beings and the beneficial organisms, contamination of agricultural products with pesticide residues, pollutuion of soil and water, the potential hazards to human beings from the unintentional ingestion of of pesticides residues in food and water, disruption of non-target organisms; resurgence of pests or potential pests; and resistance of pests to pesticides. In order to combat such problems more and more new pesticides have been developed, creating the situation called pest warfare. The environmental movement of the public has given pressures to polticians, governments and industries. The situation has brought in strict regulation of pesticides in terms of environmental safety and shorter economic life-span of pesticides. An overall readjustment of pest control or plant protection strategy became urgent. Reflecting the situation, more terms relating to safety evaluation and toxicology; metabolism and degradation as well as movement and environmental fate of pesticides; non-target orgasnisms, wildlife and ecology; soil and water; resistance, tolerance and resurgence of pests have been introduced in our disctionary.

On the other hand, development of selective and biodegradative pesticides, which hopefully overcome pest resistance became an urgent need. There are several directions to increase the efficiency of development procedures: a) development of biorational design of pesticides; b) finding new lead compounds based on natural products; c) sophistication of optimizing molecules based on computer-assisted structure-activity relationships; d) development of biorational strategy and tactics for combating pest resistance; e) improvement of pesticide delivery system based on the innovation of pesticide formulation and application techniques; and f) sophistication of bio-assay methods.

The trend is emphasizing the development of rational design of pesticides, which introduced many terms relating to biochemistry and physiology of pests and crops, particularly those related to target sites, mechanism and mode of action and pesticide design like structure-activity relationships.

There are other impact on pesticide development from industrial side. Since 1945 the progress of technology development of synthetic pesticides has been remarkable and pesticide business had been one of the profitable ones and much investment has been done in the past. Pesticide development will not be achieved by a single science or technology, but requires an integration of various types of science and technology. The maintenance and development of such system requires enormous input of manpowers and money. When the market is expanding, which is enough to absorb the preliminary investment, such input is justifiable. When one technology is becoming matured, the outcome as compared with the input effort is becoming smaller. In such a case, new technologies which are not related to the old ones are emerging, invading the existing market, and eventually replacing the old ones.

What about pesticides. Pesticide technology has reached an amazing level to produce pesticides which are effective at the level of one gram per hectar or below. This again means that the newer pesticides must be more than that level. To be so, the activity will be specialized and the spectrum will be narrowed, resulting in the smaller market for the product or difficulty in finding broad spectrum products. Secondly, due to the development of analytical techniques, the safety evaluation on human body and the environments becomes more detailed. This means more time, manpower and money expenditures, and the potentially remarkable potent products often drops from development schedules.

The present pesticide technology is entering into the ripening stage from any points of view and the efficiency of the input becomes rapidly going down. This trend is discouraging some of the chemical companies and so-called Merge and Aquisition is becoming wide-spread in the U.S. and Europe. On the other hand, pesticide industries in the world have created an organisation called GIFAP(International Group of National Associations of Agrochemical manufacturers) to cooperate each other, to contact with International agencies, such as FAO, WHO, UNEP, IRPTC, and to combat with public outcry concerted with politics.

Pesticide use in under-developed and developing countries provides the problems not encountered in developed countries. There are many constrains which inhibit the proper use of pesticides in these areas, mostly socio-economic in nature. Many terms relating to economy, socio-economics, politics and international relationshios are now included in our dictionary.

What will be the breakthrough ? The problems discussed above are often interrelated, and actually have resulted from a single dependency on poisonous chemicals for pest control. There is considerable interest in expanding the integrated pest management(IPM) approach - in short - a strategy that utilizes knowledge, monitoring, action criteria, materials, and methods, in concert with natural mortality factors, to manage pest populations(The Pesticide Conpiracy, R. van den Bosch). Therefore, all the components of IPM consistes of a part of our dictionary. However, emphasis is often put on keeping pesticide use to a minimum and relying as much as possible on biological and natural controls as well as improved cultivation practices to prevent the build-up of pest population to the level which will cause economic damage.

This trend apparently seems to be antagonistic to pesticide development. However, when we change-over the way of thinking, it has the potential to refresh pesticide science and technology.

How then will this concept affect pesticide development. Insecticides have been and will be the representative for development of pesticides. Therefore, examples will be taken often from the field of insecticides.

Only when a chemical attacks a suitable weakness of insects does it cause a serious lesion, resulting in the death of the insects. Such is the way of conventional insecticides, and since such vulnerable targets are amazingly small in number despite the efforts to explore the targets, to date the type of insecticides are therefore very limited: organochlorines, organophosphorus, carbamates, pyrethroids and few others.

However, based on IPM, insecticides need not necessarily be broad in spectrum and potent as well as fatal, but serve to decrease the insect population and thus reduce the damage level.

If this can be done, the type of potential compounds will then be greatly increased and exploration for insecticides now on will be based on our understanding of physiology and ecology of pests, natural enemies and economic organisms, particularly in terms of the bio-active chemicals involved, thus promting research on basic life processes and natural products.

One fast growing area is chemical ecology, which is an ecological understanding of the origin, function, and significance of natural chemicals(ecochemicals) that mediate interactions within and between organisms(Journal of Chemical Ecology). With a deeper understanding of insect behavior, more candidate ecochemicals will be discovered. Also, such ecochenmicals are not limited to the insect world. Among crops, weeds, pest insects, natural enemies and microbes, a net work which can be called chemical communication system has been evolved. By manupulating such communication systems with ecochemicals, the prospect of controlling harmful organisms with minimum consequences on the ecosystem will become more realistic.

Also, the understanding of regulatory mechanisms within individual organisms such as hormones has been much deeper than in earlier days.

The basic concept is not only kill the pests but more importantly to manage pest population by affecting Achilles' heel of pests and to activate the vigor of economic plants and animals. The followings are examples of the targets of such approaches:

crop modification to be resistant to disease; feeding, copulation, oviposition stimulation-depression; hatching acceleration-depression; aphrodisia; attraction-repellent action; escaping; alarming; defence; trail following; aggregation; induction-inhibition of food searching; diapause breaking;

metamorphosis induction; parasitism-symbiosis; infection; caste differentiation; settelemnt stimulation-depression; alleropathy; host selection; hypersensitive reaction- depression; inhibition of pheromone production; cuticle formation; inhibition of hormone production etc. Furthermore, there are synthetic chemicals analogous in function to natural ones, which are not necessarily similar to the natural ones in their chemical structures.

Pesticide Science is an umbrella to integrate all the individual science like chemistry, biochemistry, entomology, plant pathology, weed and crop sciences, ecology, toxicology and environmental science in order to develop pesticides and its safe and efficient use. However, pesticides in Englsh implies biocides, not very proper for the ideal types of chemicals which have been discussed. Pesticstatics or pest control chemicals can cover wider range of chemicals but excluding plant growth regulators. The word " NOYAKU" in Japanese means agro-drugs, corresponding to " Agrochemicals". Although NOYAKU or Agrochemicals has an advantage to cover also plant growth regulators, which are surely not pest-cides, the problem associated with these terminology is to limit the targets in agricultural sect. Actually so-called pesticides covers to control public health and household insects, mites and ticks, fungal and bacterial diseases of any stored produce and fabrics(usually medicinal and veterinary fungicides as well as industrial bactericides and disinfectants, anti-microbial drugs for animal and man are excluded), weeds not related to agriculture, endoparasites of animal, rodents in cities, molluscs not related to agricultur, etc.

Use of the word "bioregulators" may initiate changes in awareness and create new trends among academia, industry and public. The study and application of these agents are sure a part of biotechnology.

Quantities and Units in Biotechnology

Kozo Kuchitsu

Department of Chemistry, Nagaoka University of Technology
Nagaoka 940-21 Japan

1. INTRODUCTION

The terminology, symbols and units of physical quantities constitute a major part of "the grammar of science and technology", by which one can acquire, record, transmit, and digest useful scientific information [1]. Since biotechnology is one of the truly interdisciplinary and international fields of research, it is essential to establish "a standard language of communication" for description of information that can be handled easily and understood correctly by anybody, anywhere, and at any time. Thus proper use of units and symbols of physical quantities is just as important as that of chemical and biological nomenclature.

The present chapter aims to summarize the current activities of international organizations for standardization and to abstract their recommendations in regard to "the basic grammar". Since the author has been associated with IUPAC (International Union of Pure and Applied Chemistry) Commission I.1 (Physicochemical Symbols, Terminology and Units) and IDCNS (Interdivisional Committee on Nomenclature and Symbols), most examples are taken from those in physical chemistry. Nevertheless, the recommendations are so general that they are equally applicable to all fields of chemistry, biochemistry and biology, as well as any other academic fields.

2. INTERNATIONAL ACTIVITIES FOR STANDARDIZATION

The International System of Units (SI) was adopted by the 11th General Conference on Weights and Measures (CGPM) in 1960. SI has since been accepted in all areas of human activities and in almost all the geographic regions of the world [2]. CGPM has the ultimate authority on the resolutions and recommendations regarding units; additions and amendments are made at each conference.

K. Loening (Editor)
The Terminology of Biotechnology:
A Multidisciplinary Problem

The formal resolutions adopted by CGPM concerning the general principles on units are taken into the recommendations on the practical use of these units in our daily work by ISO (International Organization for Standardization). Its Technical Committee 12 (ISO/TC12) publishes a wide variety of recommendations on the symbols and units of physical quantities, as well as mathematical symbols for use in physical sciences and technology [3].

In the field of chemistry, IUPAC takes the initiative [4]. The function of Commission I.1 is "to promote uniform practice by the chemical community in the usage of symbols, terminology, and units and to assure proper coordination with other IUPAC bodies and international scientific organizations in all matters related to symbols and units [4]." The first "Green Book" was published in 1969, and the second and third revised editions were published in 1973 and 1979, respectively [5]. A working party established in this Commission in 1983 published a completely revised and extended version in 1988 [6,7]. This new Green Book, ix+134 pp., attempts to provide a manual of advice and assistance for the day-to-day use of practicing scientists. Its principal part is the tables of recommended names and symbols for the physical quantities, with their SI units and definitions*, most likely to be used by chemists. An example of the tables is given in Table 1.

Table 1 Extract from the Green Book [6]: "General Chemistry"

Name	*Symbol*	*Definition*	*SI unit*	*Notes*
amount (of substance)	n	$n_B = N_B/L$	mol	1, 2
Avogadro constant	L, N_A		mol^{-1}	
mass of atom, atomic mass	m_a, m		kg	
molar mass	M	$M_B = m/n_B$	kg mol^{-1}	2, 4
molar volume	V_m	$V_{m,B} = V/n_B$	m^3 mol^{-1}	2, 4
mass fraction	w	$w_B = m_B/\Sigma m_i$	1	2

(1) The words 'of substance' may be replaced by the specification of the entity.

Example When the amount of O_2 is equal to 3 moles, $n(O_2) = 3$ mol, then the amount of $½O_2$ is equal to 6 moles, $n(½O_2) = 6$ mol. Thus $n(½O_2) = 2n(O_2)$. See also p.41 below.

(2) The definition applies to entities B which should always be indicated by a subscript or in parentheses, e.g. n_B or $n(B)$.

(4) The definition applies to pure substance, where m is the total mass and V is the total volume. However, corresponding quantities may also be defined for a mixture as m/n and V/n, where $n = \Sigma n_i$. These quantities are called the mean molar mass and the mean molar volume respectively.

(*In most cases the definition of each physical quantity is simply a one-line equation, but the definition should be useful for identification purposes. More complete and formal definitions of many technical terms and physical quantities are to be found in the Gold Book [8].) The contents of the Green Book have been made fully consistent with the CGPM and ISO documents [2,3] mentioned above. An abbreviated list of the Green Book was also published [9]. This laminated card of four A4-size pages provides a ready reference to the symbols and units of quantities, fundamental physical constants, and other essential material most often used by authors, teachers and students in chemistry and related disciplines.

A number of commissions in other divisions of IUPAC are also working actively for standardization in specific disciplines. In particular, Commission VI.2 (Commission on Biotechnology, Applied Chemistry Division) has recently completed "Selection of Terms, Symbols and Units Related to Microbial Processes" [10]. This document is a revised version of its predecessor, "List of Symbols with Units Recommended for Use in Biotechnology [11]". The new version [10] includes some clarifications and additions of new quantities which describe particular characteristics of growth and product formation, specifically, integral energy characteristics. In addition, an adjustment has been made to conform to the recommendations made in the new Green Book [6]. The new list of symbols and units [10] consists of four sections: (1) General quantities such as molar activation energy, pressure, temperature, and yield, (2) Concentrations and amounts, (3) Intensive quantities such as density and viscosity, and (4) Rate quantities of various kinds.

Two valuable recommendations related to biotechnology have been published by the IUPAC/IUPAB/IUB Interunion Commission on Biothermodynamics (IUPAB: International Union of Pure and Applied Biophysics, IUB: International Union of Biochemistry) for measurement and presentation of biochemical equilibrium data [12] and presentation of thermodynamic and related data in biology [13]. The former document is composed of a set of standard conditions for equilibrium measurements and a set of recommended terminology, symbols and units. The latter document includes (1) a list of related physical quantities, units and symbols, (2) some comments concerning choice of units in biothermodynamics, (3) notations for variables, states and processes, and (4) those related to biological solutions.

3. PHYSICAL QUANTITIES AND THEIR SYMBOLS

Quantity calculus: When one measures a physical quantity, e.g., the wavelength of a certain radiation, one takes a known reference standard that has the same dimension (length) and measures the ratio of these quantities. This reference is called the unit. Therefore, the value of a physical quantity is expressed as

$$\text{physical quantity} = \text{numerical value} \times \text{unit}. \quad (1)$$

For instance, the wavelength of one of the sodium-D emission lines can be represented as

$$\lambda = 589.6 \text{ nm} \quad (2a) \qquad \text{or} \qquad \lambda/\text{nm} = 589.6 \quad (2b)$$

Neither the physical quantity, nor the symbol used to denote it, should imply a particular choice of unit. Thus, one is *not* allowed to say: "λ = 589.6, where λ denotes the wavelength in nm."

Equation 1 can be used in multiplications and divisions of physical quantities, where numerical values and units can be handled simultaneously by normal algebra. This scheme, called *quantity calculus,* is particularly convenient in the conversion of units (see below) and in displaying the numerical values of physical quantities in tables or in figure axes (Table 1 and Fig. 1) [6].

Table 1 Example of table headings

T/K	10^3K/T	p/MPa	ln(p/MPa)
216.55	4.6179	0.5180	−0.6578
273.15	3.6610	3.4853	1.2486
304.19	3.2874	7.3815	1.9990

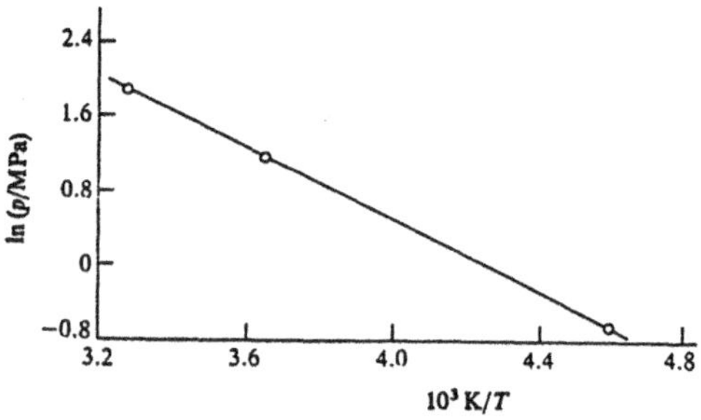

Fig. 1 Example of figure axes

Base and derived quantities: By international convention physical quantities are organized in a dimensional system built upon seven "base quantities", each of which is regarded as having its own dimension [2]. These base quantities and the symbols used to denote them are listed in Table 2. All other physical quantities are called "derived quantities" and are regarded as having dimensions derived from the base quantities by multiplication or division. The relationship between the base quantities and derived quantities correspond to that between the base units and derived units (see section 4).

Table 2 Names and symbols of SI base quantities and units

SI Base quantity	Symbol	SI base unit	Symbol
length	l	metre	m
mass	m	kilogram	kg
time	t	second	s
electric current	I	ampere	A
thermodynamic temperature	T	kelvin	K
amount of substance	n	mole	mol
luminous intensity	I_v	candela	cd

Symbols for physical quantities: The symbols for a physical quantity should generally be a single letter of the Latin or Greek alphabet, either capital or lower case and printed in *italic* type. All other symbols are printed in roman type. It may be modified by subscript and/or superscript of specified meaning, or further characterized when necessary through annotations in parentheses put directly behind the symbol. Subscripts and superscripts that are themselves symbols for physical quantities or numbers are printed in *italic* type, but in all other cases in roman type (Examples: C_p for heat capacity at constant pressure, but V_m for molar volume).

Names and symbols of physical quantities: Extensive recommendations have been made on the names, definitions and symbols of physical quantities [3,6,8,10,12,13]. Unlike the names, definitions and symbols of units, which are adopted by international convention and are mandatory [2], those of physical quantities are "recommendations" chosen to conform to current usage and to minimize conflicts so far as possible. Therefore, authors are free to choose any symbols they wish for the quantities they discuss, but each symbol should be defined unambiguously and used consistently in order to avoid any possible confusion; if the authors inevitably make major deviation from the recommended symbols, they must specify their notations with particular caution.

Examples of physical quantities: The physical quantity "amount of substance" is of special importance to us. It is proportional to the number of specified elementary entities of that substance, the proportionality factor being the same for all substances; its reciprocal is the Avogadro constant, $L = 6.022 \times 10^{23}$ mol^{-1} (not the Avogadro *number*, because it has a dimension of mol^{-1}.) The SI unit of this quantity is the mole. The "amount of substance" was formerly called "number of moles", but this name should no longer be used, just as the physical quantity "mass" is never called, with one of its units, "number of kilograms". The name "amount of substance" may be abbreviated to "amount" in such cases as "amount of N_2" and "amount concentration". The elementary entities may be chosen as convenient, not necessarily as physically real individual particles. It is essential to specify the entities to avoid ambiguities: e.g., amount of Cl is denoted as $n(\mathrm{Cl})$; amount of Cl_2 is denoted as $n(\mathrm{Cl_2})$; 1.5 moles of Hg_2Cl_2 have a mass of 708.13 g.

The Green Book defines two adjectives: "specific" and "molar". An extensive quantity, i. e., a quantity proportional to the mass or amount of substance of a system, is often symbolized by a capital letter. The corresponding *specific* quantity, i. e., quantity divided by mass, may be represented by the corresponding lower case letter: e. g., volume V and specific volume $v = V/m$. A subscript m to the symbol of an extensive quantity denotes a *molar* quantity i. e., quantity divided by amount of substance: e. g., molar volume $V_\mathrm{m} = V/n$. It is permissible to omit the subscript m when the text makes it obvious that a molar quantity is intended. There are, however, a few cases where the adjective "molar" has a different meaning, namely divided by amount concentration: e. g., molar absorption coefficient and molar conductivity.

The symbols used to represent the states of aggregation of chemical species, as appendices to symbols of a chemical formula or a physical quantity, and those in chemical thermodynamics, used either as superscripts (to denote the states or conditions) or subscripts (to denote a chemical process or reaction) are listed in Table 3.

4. UNITS AND THEIR SYMBOLS

International System of Units: The International System of Units (SI) is a coherent system of units built from seven SI base units, one for each of the seven dimensionally independent base quantities (Table 2) [2]. The SI derived units are expressed as products of powers of the base units with numerical factors equal to unity. The units for plane angle and solid angle, radian (symbol rad) and steradian (symbol sr), are described as SI supplementary units [2], but they may also be treated as dimensionless derived units. Hence, rad and sr may either be used explicitly as units when ap-

Table 3 Symbols used to specify the states of chemical species

Appended to a chemical formula or a physical quantity in parentheses				Used as superscript	
g	gas or vapor	vit	vitreous substance	⊖,o	standard
l	liquid	a,ads	species adsorbed on a substrate	*	pure substance
s	solid	mon	monomeric form	∞	infinite dilution
cd	condensed phase (solid or liquid)	pol	polymeric form	‡	activated complex, transition state
fl	fluid phase (gas or liquid)	sln	solution	E	excess quantity
cr	crystalline	aq	aqueous solution		
lc	liquid crystal	am	amorphous solid		

Used as subscript to a symbol in chemical thermodynamics			
vap	vaporization, evaporation (liquid - gas)	ads	adsorption
sub	sublimation (solid - gas)	dpl	displacement
fus	melting, fusion (solid - liquid)	imm	immersion
trs	transition (between two phases)	r	reaction in general
mix	mixing of fluids	at	atomization
sol	solution (of solute in solvent)	c	combustion reaction
dil	dilution (of a solution)	f	formation reaction

propriate or omitted if clarity is not lost thereby.

Symbols for units should be printed in roman type. They should remain unaltered in the plural and should not be followed by a full stop except at the end of a sentence. They should be printed in lower-case letters, unless they are derived from a personal name, when they should begin with a capital letter, e. g., Hz for hertz (SI derived unit of frequency), N for newton (force), Pa for pascal (pressure), and J for joule (energy). An exception is the symbol for the liter (a non-SI unit used together with the SI units), which may be either L or l, to avoid confusion with numeral 1.

Any of the approved decimal prefixes, called SI prefixes, may be used to construct decimal multiples or submultiples of SI units (Table 4). Only one prefix symbol may be used without space preceding the unit symbol (except for °C and kg), and the combination

Table 4 SI Prefixes

Submultiple	Prefix	Symbol	Multiple	Prefix	Symbol
10^{-1}	deci	d	10^{1}	deca	da
10^{-2}	centi	c	10^{2}	hecto	h
10^{-3}	milli	m	10^{3}	kilo	k
10^{-6}	micro	μ	10^{6}	mega	M
10^{-9}	nano	n	10^{9}	giga	G
10^{-12}	pico	p	10^{12}	tera	T
10^{-15}	femto	f	10^{15}	peta	P
10^{-18}	atto	a	10^{18}	exa	E

is taken as one symbol, e. g., 1 ms^{-1} = 10^{3} s^{-1}.

In general there is only one SI unit for each physical quantity. However, the reverse is not true. Several physical quantities may have the same SI unit; for example, the SI units of heat capacity and entropy are both J K^{-1}. In exceptional cases physical quantities with the same dimension may have different SI units and/or unit symbols. For example, the relationship between the Celsius temperature, θ, with SI unit °C (degree Celsius), and the thermodynamic temperature, T, with SI unit K, is defined by

$$\theta/°C = T/K - 273.15 \quad (3)$$

A common unit for amount concentration is mol dm^{-3}. A solution of e. g., 1 mol dm^{-3} is often called a one molar solution and denoted as 1 M solution. However, M should not be treated as a symbol for the unit mol dm^{-3}, in the sense that it should not be used with SI prefixes or in conjunction with other units, e. g., mM or M s^{-1}.

Non-SI Units: It is recommended that only SI units be used in science and technology (with SI prefixes where appropriate). Where there are special reasons for making an exception to this rule, it is recommended always to define the units used in terms of SI. Almost all non-SI units are now defined precisely* in terms of the corresponding SI units: e. g., thermochemical calorie, cal_{th} = 4.184 J; atmosphere, atm = 101 325 Pa; torr = atm/760 ≈ 133.322 Pa; conventional millimeter of mercury, mmHg = 13.5951 × 980.665 × 10^{-2} Pa ≈ 133.322 Pa; and poise, P = 0.1 Pa s. Note, however, tha these non-SI units should be replaced by the corresponding SI units. (*The values of two non-SI units, eV (electronvolt) and u (unified atomic mass unit) in terms of the corresponding SI units depend on the fundamental physical constants, which are determined by experiment.) The unit u is often called the dalton, with symbol Da, although the name and symbol have not been approved by CGPM. Other common non-SI units used with the SI are min for minute, h for hour, d for day, ° for degree, ′ for minute (plane angle), and ″ for second (plane angle).

Conversion of units: Conversion among the SI units is straightforward, because all SI units are coherent. Quantity calculus may be used to make conversion of non-SI units used in older literature. For example, the vapor pressure of water at 20 °C in torr may be converted to that in atm by use of the definitions of these non-SI units given above:

$$p(\mathrm{H_2O},\ 20\ ^\circ\mathrm{C}) = 17.5\ \mathrm{torr} = 17.5 \times 133.3\ \mathrm{Pa}$$
$$= 17.5 \times 133.3\ (\mathrm{atm}/1.013 \times 10^5) = 2.30 \times 10^{-2}\ \mathrm{atm}$$

5. MATHEMATICAL SYMBOLS, ABBREVIATIONS AND ACRONYMS

Numbers should in general be printed in roman type. The decimal sign between digits in a number should be a point or a comma. To facilitate the reading of long numbers the digits may be grouped in threes about the decimal sign but no point or comma should be used except for the decimal sign. A number smaller than unity should always have a zero preceding the decimal sign.

Example: 2 573.421 73 or 2 573,421 73 or 0.2573 *not* .2573

Letter symbols for mathematical constants (e. g., e, π, i) should be printed in roman type, but letter symbols for numbers other than constants should be printed in *italic* type. Likewise, symbols for special mathematical functions (e. g., log, exp, sin, d, Δ) should be printed in roman type, but symbols for a general function (e. g., $f(x)$) should be printed in *italic* type. Vectors and matrices should be printed in ***bold face italic*** type. Reference [3] lists many mathematical symbols for use in science and technology.

Abbreviations and acronyms (words formed from the initial letters of groups of words that are frequently repeated) should be used sparingly [14]. Unless they are well established (e. g., DNA) they should always be defined once in any paper, and they should generally be avoided in titles and abstracts. Abbreviations used to denote physical quantities should, if possible, be replaced by the recommended symbol for the quantity (e. g., E_i rather than I.P. for ionization energy).

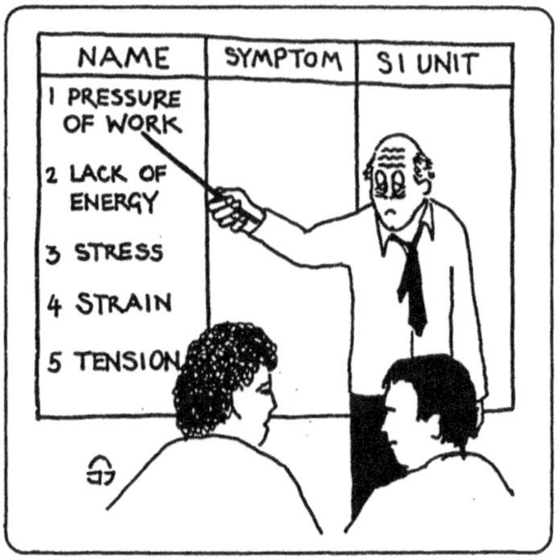

Fig. 2 Taken from [7], page 4

6. SUMMARY

1) Use recommended terminology and symbols for physical quantities whenever possible.
2) Give clear definitions of all symbols for the physical quantities used.
3) Use SI units whenever possible. In case non-SI units are used, define them clearly in terms of the corresponding SI units.

7. REFERENCES

1 McGlashan ML (1971) Physicochemical Quantities and Units. The Grammar and Spelling of Physical Chemistry, 2nd ed, The Royal Institute of Chemistry, London
2 International Bureau of Weights and Measures (1985) International System of Units (SI), 5th French and English ed, BIPM, Sèvres
3 International Organization for Standardization (ISO)(1982) ISO Standards Handbook 2, Units of Measurement, ISO Central Secretariat, Geneva
4 IUPAC Secretariat (1988) IUPAC Handbook 1987-1989 (The Blue Book), Blackwell, Oxford
5 Whiffen DH (1979) Manual of Symbols and Terminology for Physicochemical Quantities and Units, 3rd ed. Pure Appl Chem 51:1
6 Mills IM, Cvitaš T, Homann KH, Kallay N, Kuchitsu K (1988) Quantities, Units and Symbols in Physical Chemistry (The Green Book), Blackwell, Oxford
7 Kuchitsu K (1989) Chem Int 11:1
8 Gold V, Loening KL, McNaught AD, Sehmi P (1987) Compendium of Chemical Terminology: IUPAC Recommendations (The Gold Book), Blackwell, Oxford
9 Homann KH (1987) Abbreviated List of Quantities, Units and Symbols in Physical Chemistry, Blackwell, Oxford
10 IUPAC Secretariat (1989) Chem Int 11:239 announcing Eroshin VK, Selection of Terms, Symbols and Units Related to Microbial Processes, Pure Appl Chem (to appear)
11 Righelato RC (1982) Pure Appl Chem 54:1743
12 Wadsö I (1977) Biochem Biophys Acta 461:1
13 Wadsö I (1986) Pure Appl Chem 58:1405
14 Lide DR, Jr (1980) Pure Appl Chem 52:2229

Engineering Terminological Concerns in Biotechnology

Murray Moo-Young
Department of Chemical Engineering, University of Waterloo
Waterloo, Ontario, Canada N2L 3G1

Abstract: The present state of evolution of biotechnology as a multidisciplinary field has revealed concerns for real engineering expertise for implementation of the various relevant biology-based techniques. Technical language difficulties are beginning to become apparent. The need to address them by the collective deliberations of professional organizations traditionally devoted to related but separate disciplines is clear.

BACKGROUND

In 1984, the Office of Technology Assessment, US Congress, came to the conclusion that "the commercial success of industrial applications of recombinant DNA and cell fusion techniques will hinge on advances in bioprocess engineering"(1). As modern biotechnology, the so-called "new" biotechnology, evolves from the development of genetic manipulative techniques and methodologies to potential manufacturing feasibilities, the real engineering aspects of the technology become increasingly important(1, 2). Indeed, it is engineering or the lack of it which seems to be responsible for the historical trend of "boom or bust in the biobusiness"(3).

At present, the under-developed state of the language of biotechnology in a comprehensive sense adds to the confusion in efforts for collaborative team work which is essential for commercial implementation of a technology. Here, we draw attention to some of the dilemma which relate to terminologies, nomenclature for symbols and units of quantification as examples of a potential scenario for a biotechnology "tower of Babel" for engineers.

ENGINEERING IN BIOTECHNOLOGY

As a reminder, Figure 1 illustrates the multidisciplinary nature of biotechnology in which the biological/life sciences, chemistry and engineering are essential disciplinary underpinnings but can, in effect, represent "three solitudes".

K. Loening (Editor)
The Terminology of Biotechnology:
A Multidisciplinary Problem

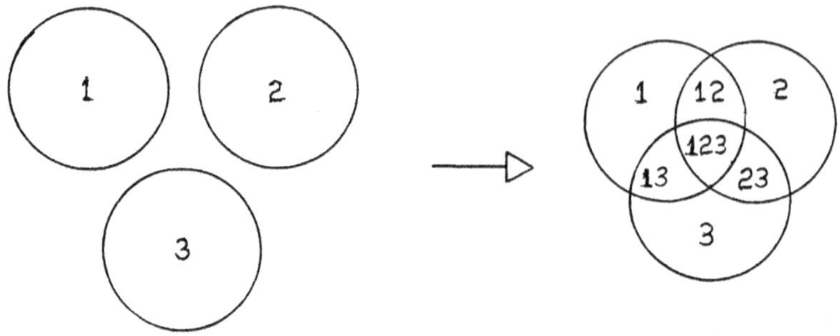

<u>Figure 1:</u> Roots of the Principles of Biotechnology (1-2-3) as a Multidisciplinary Field: (1) Biology, (2) Chemistry, (3) Engineering; (1-2) Biochemistry, (1-3) Biophysics, (2-3) Chemical Engineering; (1-2-3) Biotechnology.

Historically, the biological sciences have played a dominant role in the development of biotechnology, old and new. This is understandable in view of the advances in the development of new and improved biological agents for bioconversion and biotransformation processes and the socio-economic relevance of the user sector of biotechnology products especially in medical/healthcare applications. However, as illustrated in Figure 2, the need for appropriate integration of the biological, chemical and engineering sciences is evident for synergistic modalities (4).

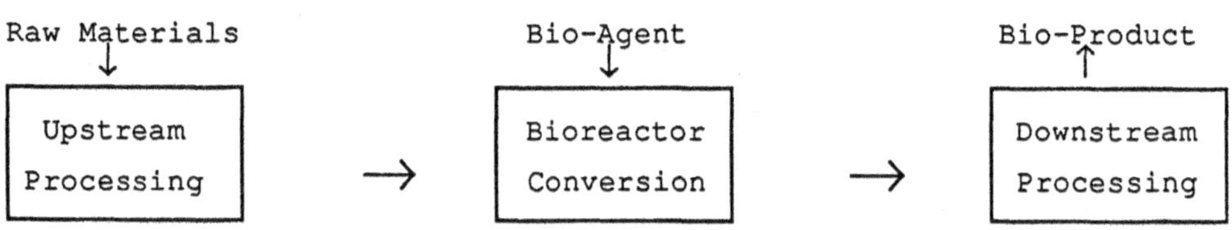

<u>Figure 2</u>. Advances in the Development of New/Improved Bio-Agents and/or Bio-Products Tend to Mask Importance of Engineering in Bio-Process Implementation.

The proliferation of specialized branches of biotechnology is indicative of the vastness of the fluid. Figure 3 attempts to illustrate the one commonality of the engineering-based branch of industrial biotechnology among the other three generic branches of microbial, animal and plant biotechnologies as application venues.

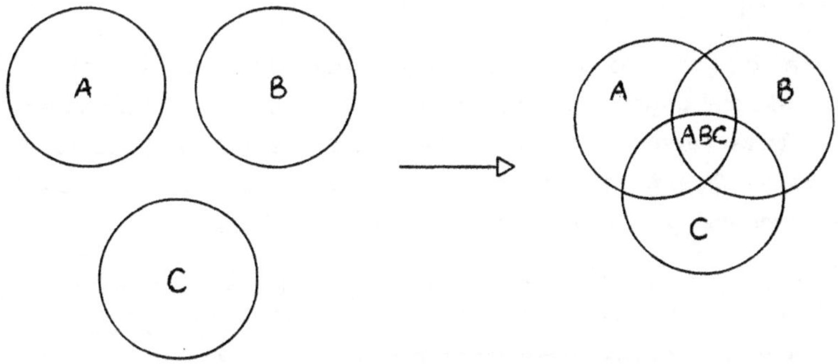

Figure 3. Commonality of Engineering-Based Industrial Biotechnology (ABC) Among Other Applications Specializations of (A) Microbial (B) Animal and (C) Plant Biotechnologies.

TERMINOLOGIES

As a reflection of the "solitudes" mentioned above, most publications of "glossaries of terms" usually have a biological/life sciences bias and little, if any, attention is given to the engineering terms. One of the few attempts at a comprehensive glossary was done in 1985(4) but it is in need of updating. The seriousness of this omission is compounded by the somewhat cavalier use of the word engineering as in "genetic engineering" for techniques. Table 1 highlights the common abuses which, in principle, could be challenged by professional engineering associations in several countries including Canada and the USA which act as licensing authorities for practicing engineers in a similar manner as "doctors" practice medicine.

Table 1: Some Pseudo and Real Engineering Terms Currently Used in Biotechnology Jargon.

PSEUDO	REAL
Genetic Engineering	Bioprocess Engineering
Protein Engineering	Biochemical Engineering
Bimolecular Engineering	Fermentation Engineering
Biomimetic Engineering	Enzyme Engineering

Pseudo-engineering terminologies appear to be proliferating as new techniques, tools and methodologies are developed in biotechnology.

NOMENCLATURE AND SYMBOLS

Because of the early independent development of the various "cultures" which generate biotechnology, we find instances where different terms are used for the same variables/parameters and different variables/parameters for the same meanings. As before, we cite some examples in Table 2 to illustrate the point.

Table 2: Examples of Different/Conflicting Nomenclature and Symbols Currently Used in the Biotechnology Literature.

ENGINEERING ROOT	**NON-ENGINEERING ROOT**	**VARIABLE OR PARAMETER**
Space Velocity (S)	Dilution Rate (D)	-
CSTR	Chemostat	-
μ	η	Viscosity
(μ)	μ	Growth Rate

Several attempts have been made to standardize the nomenclature for biotechnology with due consideration given to engineering involvement (5); however, there is still no universal agreement as to which standards to adopt.

QUANTIFICATION AND UNITS

The biological/life sciences are primarily descriptive while engineering and the applied sciences tend to be more quantitative. The units of quantification, preferably SI, are important to avoid errors in engineering calculations. An attempt at comprehensive treatment of this area is to be found in Reference 4.

Often in the biological literature, the units of variables are not given or are incomplete, a situation which leads to a great deal of frustration for engineers in biotechnology process design. A notorious example of incompleteness of information is the frequent lack of a time consideration in the expression of enzyme activities in which the substrate utilization kinetics is non-linear as is often the case. A particularly confusing picture is found for cellulase enzymes whereby it is virtually impossible to rationally compare most of the published results quantitatively.

CONCLUDING REMARKS

Biotechnology, like any form of technology, needs engineers for its commercial applications (6). However, the embryonic language of biotechnology lacks consistency and quantification, prerequisites for engineering translations of basic laboratory discoveries into industrial practice. An appeal is made for improved semantics and standardization of the terminologies, nomenclature and quantification units of biotechnology as an integrated multidisciplinary field.

References:

1 Office of Technology Assessment (1984). Commercial Biotechnology: An International Analysis, US Congress, 612 pp

2 Moo-Young M (1987). In "Encyclopedia of Science and Technology", McGraw-Hill, Vol 2, 513-515

3 Yanchinski S (1987). New Scientist, 113, 44-47

4 Moo-Young M et al (eds) (1985). Comprehensive Biotechnology, Pergamon Press, Oxford, 3760 pp

5 Atkinson B (1974). Biochemical Reactors, Pion Press, London, 376 pp

6 Humphrey A E (1984). Chem Eng Prog, 80, 7-12

On the Conceptual System and Information Management: A Step Towards the Unification of Terminology in the Subject Field of Biotechnology

Helmi Sonneveld

Topterm C.V., European Division
W.G. Plein 278, 1054 SE Amsterdam

Abstract: Modern technology is changing the world in many ways. An important result is our ability to communicate within seconds from every corner of the earth. Thus, more and more importance is now attached to the exchange of information. This growing need for information demands that the data to be retrieved are systematized in such a way that the required information can be retrieved quickly, easily and relatively cheaply. Many on-line databanks exist which can provide the public with many different types of information. In practice however, the retrieval of information from these systems often meets with considerable problems. The possibilities of modern technology and the exponentional growth of the quantity of available information lead among other things to great difficulties in communication. In order to be able to systematize information to be entered into a database, the vocabulary of the subject fields in question needs to be controlled. In biotechnology there is a great need for institutional development of information resources. What is needed is a vocabulary development effort that keeps up with the developments in biotechnology to create an internationally recognized and consistent vocabulary. Assistance in this matter is given by the principles and methods of terminology and computerized terminography.

K. Loening (Editor)
The Terminology of Biotechnology:
A Multidisciplinary Problem

efficient communication not only between biotechnologists themselves, but also between biotechnologists and the general public. This development exerts a negative influence on the flow of information. The control over the existing vocabularies in the field of biotechnology and the application of unified methods in term formation for new concepts is therefore an urgent need.

What is "Terminology"

Unfortunately, the term "terminology" as used by terminologists is actually a misnomer, since terminology is not the theory of terms. Rather, the term "terminology" is assigned to at least three different concepts:

1. terminology science; terminology: inter- and transdisciplinary field of knowledge dealing with concepts and their representations (terms, symbols, etc.);
2. aggregate of terms, which represent the system of concepts of an individual subject field;
3. publication in which the system of concepts of a subject field is represented by terms [2].

Terminology work thus spans a wide area of activities, i.e. the collection and analysis of concepts and terms, elaboration of systems of concepts and of definitions, assigning of terms to concepts, etc..

The Importance of Terminological Methods and Principles as opposed to the Traditional Lexicographical Methods and Principles

In the lexicographical tradition there is usually not any sharp differentiation of the individual concept from neighbouring concepts. In strict terminology work, however, every concept is clearly differentiated from

GENERAL INTRODUCTION

The Importance of a Uniform Terminology for Biotechnology

The subject field of biotechnology is rapidly growing and multidisciplinary. Furthermore it may be observed that the quantity of available information in this subject field is constantly increasing and this increase is an exponential one. This development is already leading to great difficulties in communication: "Current vocabulary in biotechnology is suboptimal, because of the rapid growth and multidisciplinary nature of biotechnology and because of insufficient vocabulary control by authors, editors, and publishers; there is also a need for institutional development of biotechnology information resources. Consensus is lacking among users regarding the definitions of current words, and new phenomena are named inconsistently. Fabricated terms sometimes become accepted, such as "exon" and "intron", which are now in standard usage in place of the more descriptive "intervening sequences" and "coding sequences". Biotechnology vocabulary has lost some of the specificity of the Latin-based medical vocabulary; English words can be imprecise, especially if they are not carefully defined. In summary, increasing use of inconsistent or sloppy and uncontrolled vocabulary is outstripping efforts at standardization" [1].

At present, as a corollary to the rapid development in the field of biotechology, a large number of concepts has come into existence and many new concepts come into being every day. All these concepts have to be expressed by terms. Language has a restricted word and word element stock available for the formation of terms. Considering the present rate of concept formation, the limits to terminological growth will be reached very soon, i.e. the limits for assigning unambiguous terms to new concepts. Unambiguity, however, is the basis for other concepts of the same subject field. The basis of a terminological vocabulary is the conceptual system which (especially in its graphic form, see annexes) clearly shows the relationships between the respective concepts, as well as occurrences of synonymy, polysemy, and homonymy, gaps, inconsistencies and ambiguities, and which gives an instant view of the way definitions and terms in the vocabulary are formed. The description of the concepts is done by means of a definition of the concept.

In the following paragraphs I intend to present to you an example of a terminology project in progress that is based on the methods and principles of the so-called conceptual approach. I want to emphasize, however, that it is NOT my intention to declare that this approach will bring universal (terminological) happiness.
Every situation in which terminology work is done or needs to be done, whether it is translation oriented (bi- or multilingual) or not, is different and has its own characteristics. Each situation is therefore in need of its own specific analysis and solutions. For example, I have not found two translation departments, even within the Netherlands, that had the same infrastructure and had to cope with the same (organisational) problems.

The type of project approach that is presented below is interesting because it is a try-out of several ideas that have been developed within Topterm (Terminology-bureau in Amsterdam and Columbus,Ohio, U.S.A.) during the past two years and that elaborate on the theory of Eugen Wüster [3]. A description of some of the basic ideas behind the arguments in favour of the conceptual approach can be re-read in many articles published by Infoterm in Vienna.

DESCRIPTION OF THE PROJECT

The terminology project under discussion is being developed under the authority of the International Union for Pure and Applied Chemistry (IUPAC), in particular of the Subcommittee on Surface Analysis of the Commission on Microchemical Techniques and Trace Analysis (V.2) in co-operation with the ASTM (American Society for Testing and Materials) Committee E42.

The aim of the project is the compilation of a comprehensive list of terms for the subject field of Auger Electron Spectroscopy, which is a specialized field of surface analysis. The subject field is not directly related to biotechnology, but the project may be regarded as an example of method. The list of terms is to be published internationally in the English language.

Experts in the field of surface analysis some time ago came to the conclusion that there is little agreement on the meaning of the concepts used within the field, a situation identical to the situation in the subject field of biotechnology described earlier.

With regard to unambiguous communication, it is of the greatest importance that a scientific consensus is reached on the meanings of the concepts within the subject field. No one will doubt the importance, but some critical minds will certainly doubt whether such a scientific consensus can ever be reached (independently of the subject field). This, in fact, was one of the surprising experiences we had while working on the construction of the conceptual system.

In order to present an idea of the way we are approaching this project, an overview of the phases involved is given below:

1. construction of the conceptual system:
   - the delimitation of the subject field (number of terms to be added to the system)
   - the definition of the nature of the relations between the concepts
   - the classification of the concepts (from the most generic to the most specific concepts);

2. the testing of the definitions of the concepts (the contents of the concepts) in the vocabulary;
3. the testing of the terms (the linguistic forms of the concepts) of the vocabulary;
4. the final evaluation of the conceptual system (correction of cases in which synonymy, homonymy and polysemy occur);
5. the drafting up of the vocabulary for publication

Before I describe the various phases a little more extensively, I want to emphasize the fact that this project is team work. The experts in the subject field of Auger Electron Spectroscopy are working together with the terminologists of Topterm. This exchange of knowledge has already led to quite a few fruitful results. The exchange of information between scientists and linguists is one of the important aims of this pilot project.

In the course of the discussions concerning the conceptual system, the scientists started to realize how many concepts that are used daily in their work are in fact not clearly defined and understood. This resulted in discussions concerning some of the basic concepts in in which the various viewpoints were made explicit.

People look at things around them from different points of view. This is definitely also the case in science. For example, a word like plasma in biology means a certain portion of the blood, in physics it refers to superheated gases. In other words: different disciplines, different meanings. This could indeed mean that the two different points of view may cause irreconcilable differences with regard to the conceptual system.

Fortunately, the latest developments in computerized information management systems have made it possible to construct dynamic and flexible systems that are adaptable to whoever is working with the conceptual system in question and also to the prospective developments of a discipline and its specific subject knowledge. In this context one should think of, for example, expert systems and their applications. This implies that the experts adapt a more flexible attitude towards the construction of the basic conceptual system that is used for the printed version of the vocabulary. The software package is part of the final publication.

Working together clearly showed how and to what extent terminologists can contribute to the development of a conceptual system of a specific subject field.

Phase one consisted of comparing the concepts of the subject field. Of each concept, the essential characteristics were determined. On the basis of these characteristics it was determined which concepts have certain characteristics in common. The concepts with matching characteristics were gathered in one group. We finally ended up with five different groups (annex 1).

Before drawing up the conceptual system, the nature of the relationships between the concepts had to be established. A conceptual system consists of several types of relationships. First, there are relationships between the various subgroups into which the subject field can be divided. Second, there are relationships between the concepts within the subgroups. It is the task of the experts to establish the nature of the relationships between the subgroups and between the various concepts within the subgroups.

The concepts of the subject field in question are classified according to the principle that the most generic concepts (large extension, small intension) appear in the highest levels of the conceptual system, whereas the most specific concepts (small extension, large intension) appear in the lowest levels of the system (annex 2).

It must be clear to everybody that the construction of the conceptual system is the most complicated and therefore the most time-consuming phase of the project. However, I would like to stress the fact that the conceptual system in this day and age no longer needs to

serve one single purpose.
This obviously increases the cost-effectiveness and the practical use of the final (printed and computerized) product. I already referred to the software package that will accompany the printed version of the vocabulary. The conceptual system, being part of a flexible programme, may be used for other purposes within a department or organisation as well. Once the relationships between the concepts of a certain subject field have been established, the system may also be used for:

- the condensing of subject field information (abstracting);
- the search for stored subject field information (indexing, classification, thesaurus construction);
- the development of a knowledge bank and other knowledge-based systems;
- the formulation of subject field information;
- the transfer of the contents of knowledge from one language to another (interpreting, translating);
- the communication of knowledge (teaching of the subject field, transfer of knowledge) [5].

Summarized, a flexible software program makes it possible to:

- adapt the conceptual system to the developments of the subject field in question;
- adapt the conceptual system to the experts' personal vision of the subject field in question;
- to adapt to the various aims that the system may want to serve;
- to adapt to the specific situation within the organization or department;
- to adapt to the various types of information that may be obtained from the conceptual system.

Considering the possible different aims and objectives and organisational set-ups, there is little use for a rigid terminological system. Of course, this does not mean that we should lose sight of the problem of the compatibility of computerized terminological systems.

Also, this does not mean that there should be one dominating type of software package for every (terminological) situation. As has been stated earlier in this paper, every situation may require its own terminological and computerized solution. The computer system should not be allowed to dictate the organization of the terminological activity. The overall organization of a terminology project is extremely complicated.
Many factors play a decisive role, for example:

1. the internal and external organisation of the institution in which the terminology project is to be executed;
2. the organisation of the terminological data;
3. the technical (computer) possibilities;
4. the aims and objectives of the project;
5. the plans for the future.

So, before starting up a terminology project and making decisions concerning the organization of the project that cannot be reversed, experts will have to be consulted and important long-term decisions will have to be made.

Anyway, as far as our project was concerned, it proved to be a good idea to take the work "seriously", meaning that trained terminologists were working on the project full-time, instead of part-time or even in their spare-time, as often happens.

The conceptual system, as I have pointed out, can be used for more than one purpose. It is also the basis for a sound, complete and unambiguous vocabulary. The general technical dictionaries that are used for the purpose of reference or for translation activities and that are based on lexicographical methods and principles are less suited for displaying expert information concerning a subject field.

The traditional dictionaries contain several problems:

- large number of terms;
- low linguistic motivation in the formation of definitions and terms;

- possibility of the existence of the same concepts in entirely different systems of concepts in a given subject field [2].

The conceptual system solves these problems instantly. First, the construction of the system forces you to delimit the number of concepts. Only those concepts that really belong to a certain subject field remain. The meaning of a concept, established by the relationships of the concept with the other concepts, becomes evident.

The conceptual system is also the basis for the testing of the definitions and terms in the vocabulary. The principles and methods to be used in the formation of definitions and terms are explained in document 704 issued by the ISO/TC 37 Committee on terminology, entitled: "Principles and Methods of Terminology".

As far as definitions are concerned, the document states for example that "A definition should fix and if possible reflect in words the position of the given concept in the system of all related concepts to which it belongs. For this reason the characteristics should be selected in such a way that the concept is delimited against related concepts. The definitions of all concepts of a system of concepts should be consistent with one another" [5].

The conceptual system makes it possible to create definitions that are systematic, adequate, concise and complete.

In the case of terms document 704 states: "The terminology of any subject field should not be an arbitrary collection of terms, but rather a coherent terminological system corresponding to the system of concepts. Whenever possible, the terms should also reflect the characteristics of the concepts, i.e. their intension. There should be only one term for the concept named. For this reason the analysis of the existing terms and the creation of new terms should be carried out within the framework of a system of terms" [5].

The conceptual system makes it possible to create concise terms.

The last phase of the project is the evaluation of the conceptual system. After the completion of the first phases of the project one will have a complete overview of the concepts of the subject field of Auger Electron Spectroscopy. Then, the cases in which homonymy, polysemy and synonymy occur will be indicated, as well as the cases in which gaps in the conceptual system occur and need to be filled.

This project is not yet finished, but already the above described procedure has shown us some important facts, namely that

- experts are willing to co-operate with terminologists to develop the conceptual system of their subject field (they instantly see the importance of the activity), which means that they are willing to invest time in this activity;
- terminologists can show the experts how to systematize the knowledge of their specific subject field in the form of a conceptual system or of a knowledge based or expert system;
- terminological activities make experts realize how often concepts are used without a clear understanding as to their basic meaning or use, resulting in misunderstanding or even incomprehension between experts themselves and between experts and the general public;
- systematic vocabularies based on the conceptual system can be used for a number of purposes, which increases the functionality of the system (multifunctionality of the system);
- systematic vocabularies based on the conceptual system contain reliable information from both a technical and linguistical point of view;
- the flexibility of the conceptual system allows the restructuring of the system according to different scientific points of view as well as according to future changes in the subject field. The knowledge structure no longer reflects a static situation.

The conceptual system can, in summary, serve as the basis for different types of activities ranging from classification to teaching and indexing as well as the display of different points of view with regard to the contents of the various concepts and their interrelationships.

In our day and age, information plays an increasingly important role. The flexibility of the conceptual system will ascertain its importance for terminology work in particular, but definitely also for information management in general.

REFERENCES

1 Committee on Biotechnology Nomenclature and Information Organization (1986) Biotechnology: Nomenclature and Information Organization, National Academy Press: Washington, D.C.

2 Galinski Ch (1986) Terminology and information/ knowledge management, Infoterm, Vienna (13-86)

3 Wüster E (1985) Einführung in die Allgemeine Terminologielehre und Terminologische Lexikographie, Infoterm, Vienna

4 Galinski C, Nedobity W (1986) A terminological data bank as a management tool, Infoterm, Vienna

5 ISO 704 (1987) Principles and Methods of Terminology, ISO

ANNEX 1

GRAPHICAL REPRESENTATION OF THE CONCEPTUAL SYSTEM

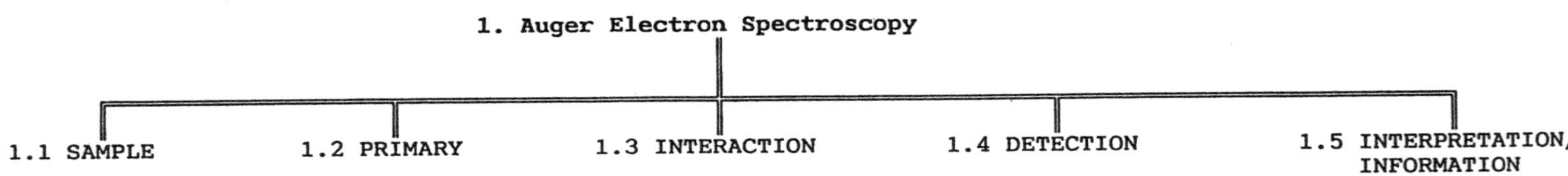

ANNEX 2

- 1.3 INTERACTION
  - 1.3.A EXCITATION
    - 1.3.A.1 single electron scattering
    - 1.3.A.2 multiple electron scattering
  - 1.3.B PRODUCTION
    - 1.3.B.1 Auger electron production (de-excitation)
      - 1.3.B.1.1 Auger process
        - 1.3.B.1.1.1 Auger electron transition (following single core-level ionization)
        - 1.3.B.1.1.2 Auger electron transition (following multiple core-level ionization and single or multiple valence-electron excitation)
        - 1.3.B.1.1.3 Auger electron transition rates
    - 1.3.B.2 secondary electron production
      - 1.3.B.2.1 secondary electrons
        - 1.3.B.2.1.1 energy distribution
        - 1.3.B.2.1.2 angular distribution
        - 1.3.B.2.1.3 secondary electron yield
  - 1.3.C AUGER ELECTRON COLLISIONS
    - 1.3.C.1 inelastic scattering
      - 1.3.C.1.1 energy loss
        - 1.3.C.1.1.1 characteristic energy losses
      - 1.3.C.1.2 inelastic scattering cross section
      - 1.3.C.1.3 inelastic mean free path
    - 1.3.C.2 elastic scattering

Computers and Biotechnical Terminology: Accessing and Integrating Data from Diverse Sources

John L. McCarthy

Information and Computing Sciences Division
Lawrence Berkeley Laboratory
University of California
Berkeley, CA 94720 U.S.A.

Abstract: Biotechnical information on computers presents special terminology challenges for locating and integrating computer data from diverse sources and levels. Standard nomenclature problems such as synonyms, homographs, and name changes are compounded when they arise on computers for names of databases, entities, and variables, as well as data values from multiple independent sources.

At the same time, computers make possible the development of new tools to meet these challenges. The data thesaurus is one such tool that helps integrate different types of data. It provides a systematic framework within which both people and computer programs can reconcile terminology from diverse, autonomous databases. Data thesauri also can help guide evolution of international standards for nomenclature and classification.

INTRODUCTION

Nomenclature is a serious obstacle to accessing and integrating our growing wealth of biotechnical data. Computers, which have fueled the data explosion, have also brought a host of new terminology challenges. This paper examines how insights and tools from computer science can help us understand and address those problems.

In order to discuss the special terminology challenges of biotechnical data, we will need to use some other terms (some might say jargon) from database management and computer science. Exhibit 1 gives a small subset of data that will introduce these terms as well as illustrate different aspects of data representation on computers.

Entities, Attributes, Values, and Domains

An ***entity*** is some "thing" about which we have data. The data in exhibit 1 pertain to a polypeptide entity named "interferon eicosapeptide." ***Attributes*** are characteristics or descriptors that pertain to an entity. In a specific computer representation, they sometimes are called fields or columns. Exhibit 1 shows three attributes: Registry, Name, and Sequence. Instances of an attribute for an entity (e.g., 79113-16-9) are called ***values***. ***Domains*** are classes or sets of permissible values for one or more attributes (such as a particular numeric range, or a set of codes -- e.g., for amino acid residues).

K. Loening (Editor)
The Terminology of Biotechnology:
A Multidisciplinary Problem

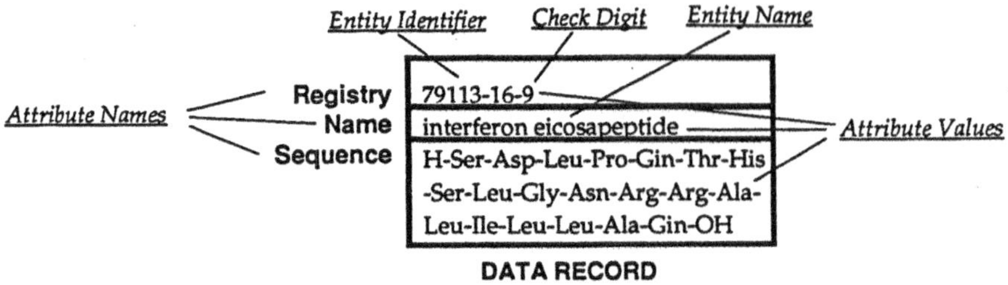

Exhibit 1: Computer Science Terms describe data components of a simple example

The terms ***type*** and ***instance*** refer respectively to general and particular aspects of other concepts -- such as entity. Thus we might say that exhibit 1 contains data for one specific *instance* of a "polypeptide" *type* of entity.

Identifiers, Names , and Data

An ***identifier*** is an attribute (sometimes called a "primary key") which unambiguously and unalterably identifies a particular instance of an entity -- such as the Chemical Abstracts Service Registration Number in exhibit 1 or a social security number for people. In order to guarantee that identifiers do not change, some experts have suggested that they have to be "meaningless" (essentially an arbitrary accession number) because any meaningful components of an identifier are subject to change [1]. To reduce transcription errors, identifiers also may contain a ***check digit*** (e.g., "9" in exhibit 1) [2; 3].

At the same time, people need ***names*** that are easy to remember -- attributes that may contain some meaningful content which can be used in place of identifiers. In the biosciences, things are frequently named to reflect biological function (such as uv123 for an ultraviolet radiation sensitive locus). Some entities may have more than one name (e.g., common name vs. formal name). Many papers in this symposium concern construction and assignment of names for specific types of entities such as proteins, carbohydrates, enzymes, interferons, monoclonal antibodies, genes, drugs, and so on.

Data is the generic term used to describe attribute values, such as the amino acid sequence in exhibit 1. Attribute values for a particular entity are frequently grouped together in physical or logical ***data record***. Values for each attribute may be constrained to a generic or special ***data type*** (e.g., integer, string, date, amino acid sequence) and to either single or multiple values (e.g., Registry vs. Sequence). In some cases (not shown in this example), attributes may themselves be complex ***data structures*** -- that is, constructs made up of other attributes. For example, an audit trail structure might consist of pairs of change dates and identifiers for persons who made each change.

Data and Metadata

Metadata is data about data, such as attribute names and types, that can be used to describe and control data values to which they pertain [4]. Metadata entities (e.g., "Attributes") in turn have their own meta-attributes, such as names, labels, synonyms, and so on. For example, the labels "Registry," "Name," and "Sequence" in exhibit 1 are metadata for attributes that describe a whole database of peptides.

DATA RETRIEVAL AND INTEGRATION REQUIREMENTS

Many people who deal with biotechnology need to *locate* relevant information and *integrate* data from different sources. These requirements have implications for the form and substance of both data and metadata. Experience from information retrieval, database management, and computer language design suggests a number of issues and guidelines.

Locating Relevant Information

There are two basic ways of using metadata to find data of interest. On the one hand, one may directly request information by specified attribute names and data values (e.g., "find Sequence values that match "ACTGGTAGCCTAAG" with fewer than 2 percent errors"). On the other hand, one may wish to browse the data or metadata at a particular level of detail (e.g., "show what attributes are available for human genes").

As the number of entities, attributes, or data values becomes large, they become difficult to locate or browse unless grouped into hierarchical classes [5]. As in a library subject catalog or thesaurus, any metadata item may participate in multiple classifications. For example, interferons can be classified as proteins (chemicals) as well as anti-infectives (drugs). The logical structure is not a simple hierarchy, but rather a directed acyclic graph, since a given node may have multiple higher level nodes.

Combining Data from Different Sources and Levels

Users want to not only browse and retrieve specified subsets of data, but also to combine and analyze data from different sources and levels. For example, one might want to identify known oncogenes on a particular human chromosome (from the Human Gene Mapping Library database), and then retrieve any available sequence information on such genes (from GenBank), along with information on related protein products (from the PIR database). Exhibit 2 (adapted from [6]) outlines current major levels of biotechnical information and databases at each of those levels.

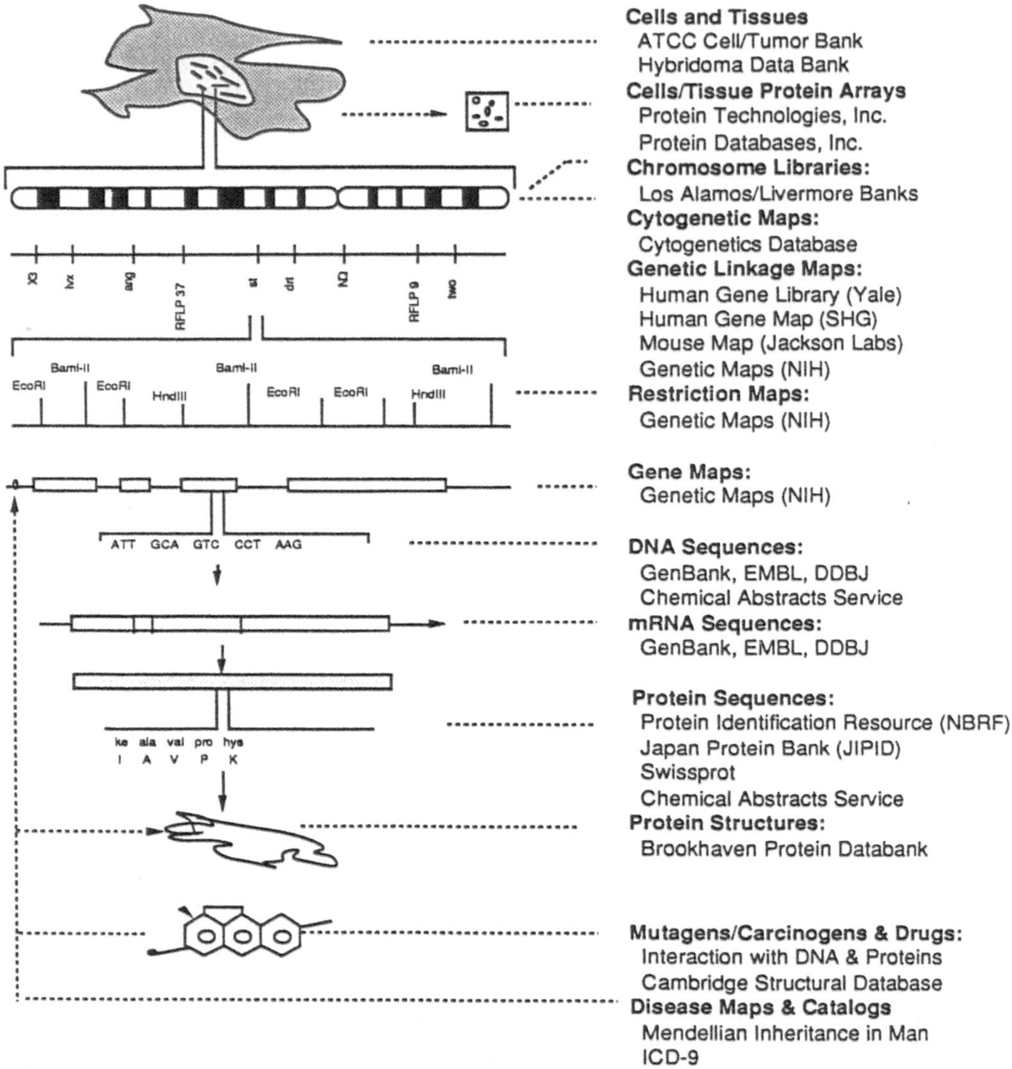

Exhibit 2: Different Types of Biological Data come from diverse databases
(adapted from "Biomedical Databases in a Universal Hierarchy of Nature" [6])

There are two basic ways to connect or link information between different databases -- direct and indirect. The most direct approach is to include identifiers from one database as attributes (foreign keys) in the other database. For example, HGML locus attributes include identifiers for literature citations (another HGML database), as well as GenBank sequences and OMIM (On-line Mendelian Inheritance in Man) entries. The indirect connection method is to match entities from different databases on the basis of attributes whose domains they share in common (e.g., gene names).

We can connect information from different databases indirectly if and only if (1) the entities in question share at least one attribute in common, and (2) the values for that common attribute can be resolved to the same domain. For each database, we must ascertain which attribute(s) may contain such information. For each such attribute, we need to know whether its domain is textual or numeric. Are values constrained to a controlled vocabulary or numeric range? What are the measurement units, if any? Finally, is the format of the attribute values fixed or"tagged" with their attribute name? Depending on the answers, it may or may not be possible to combine the information.

Consider, for example, information about physical position on a human chromosome, as represented in two major databases, GenBank and HGML, and summarized in exhibit 3.

	GenBank	HGML
Attribute Name	Location, Origin	Region
Attribute Values Domain	base pairs from Origin, relation to landmark	chromosome band
Measurement Units or Encoding	Kilobases	band number
Format/Syntax	embedded text	tag = value

Exhibit 3: Combining Data from Different Sources involves several comparability issues

Attribute Name. In GenBank, the position of an entity such as a gene is contained in a "feature table component" (attribute) whose name is Location, with measurements relative to the Origin (another attribute) of the sequence. In HGML, the comparable data element (attribute) is called Region, and its index is called Map.loc.

Value Domains. GenBank Locations are pairs of positive integers for "from" and "to/span"; in HGML Region values are cytogenetic band location codes (e.g., 21q22.3).

Measurement Type and Units. GenBank Locations are integer numbers of base pairs relative to the beginning of the sequence. The beginning position is called Origin, a text attribute with entries such as "7 bp upstream of SacI site." (Note that this "Origin" attribute is not the Origin of Replication Initiation, or ORI.) ISCN (International Standard Cytogenetic Nomenclature) band numbers in HGML are a decimal/ordinal code.

Format/Syntax. In GenBank, "from" and "to" components are separate items in a multiply ocurring data structure within the feature table. In HGML, ISCN code components (chromosome number, arm, major band, sub-band) are not separate.

Making and maintaining linkages between different databases can be done manually, on an *ad hoc* basis, or it can automated by computer software. In either case, linkage requires use of mutually agreed upon, controlled vocabularies (which can be augmented but not changed over time) for both attribute names and values, plus compatible measurement units and syntax. Terminology problems confound each aspect of data comparison.

TERMINOLOGY PROBLEMS OF COMPUTER DATA

Locating, comparing, and connecting data all face common terminology problems such as synonyms, homographs, name changes, and ill-defined specification or classification.

Synonyms, False Synonyms, and Homographs

Common terminology problems such as synonyms (different terms for the same thing), false synonyms (seemingly similar terms for different things), and homographs (the same term for different things) can assail both data and metadata. For example, the attribute named "region" in HGML sounds similar to "feature region" in GenBank, but they are in fact quite different things."Lys," "K," and "9," on the other hand, are all commonly used synonyms for data value codes in different databases for the same amino acid -- Lysine. Homographs are even more troublesome. For example, one laboratory uses the name "B1.1" for a probe that hybridizes to the MX1 locus, while another laboratory uses "B1.1" for a completely different probe that hybridizes to the APP locus.

Name Changes and Data

One of the problems with using names to connect information between databases is that names and their meaning may change over time. Some nomenclature systems such as [7] explicitly recognize that the name of any given entity may need to evolve to reflect scientific understanding of its composition or function. Such evolution includes not only simple renaming, but also differentiation (when what was originally thought to be a single entity turns out to be more than one) and consolidation (when what was originally thought to be two entities turn out to be the same) [8]. For example, the human locus associated with the disorder elliptocytosis was originally named EL. As evidence indicated that there were two loci, EL became EL1 and EL2. When specific gene products of those loci were identified, those names changed to PB41 (protein band 4.1) and SPTA (spectrin, alpha), respectively, and *EL became the allelle name [7].

Data and Metadata Specification and Classification

Information retrieval is difficult (at best) if attributes are not discrete, consistent, and well-defined. When multiple types of information are contained in a single attribute, it becomes more difficult to locate, process, and connect to related data. Data entity names are especially susceptible to information overload when people try to summarize several different characteristics in a single name, rather than as separate attributes [9]. Classifications need not follow a single hierarchy, but they must be consistent. If controlled vocabularies and classification schemes are not used for non-numeric attributes, locating such data becomes haphazard -- and linkage becomes impossible.

THE DATA THESAURUS: A NEW TOOL FOR TERMINOLOGY

At the same time as computers have posed new challenges, they also have enabled development of tools to address problems of data terminology and integration. The data thesaurus is one such software tool that deals with a number of the issues discussed in the preceding sections. Originally developed in conjunction with a prototype information system for material properties data [10; 11], the data thesaurus concept extends ideas from earlier metadata tools, including statistical codebooks, subject term thesauri, data dictionaries, and database schemas in order to:

- manage definitions and cross references for various types of metadata;
- reconcile diverse nomenclatures and classifications from multiple sources; and
- link information between different types of entities from federated databases.

General Architecture and Features

Conceptually, the data thesaurus is a logical layer that lies between the database management system (DBMS) and users or programs, as pictured in Exhibit 4. It includes a specialized database of metadata, special functions for indexing and access, and interfaces for people and software. It can be implemented using the database management system.

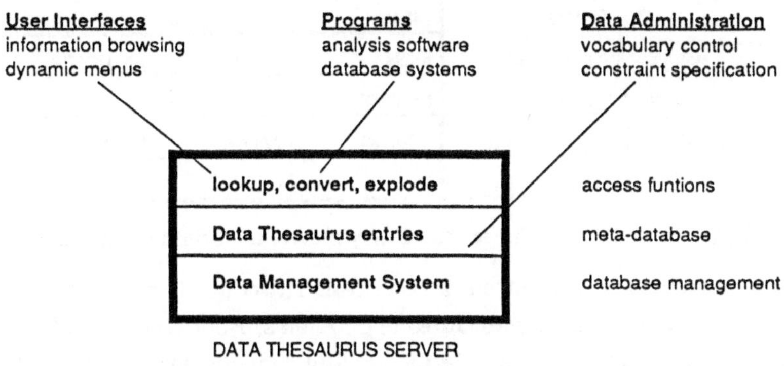

Exhibit 4: Modular Data Thesaurus Architecture supports people and computer programs

Thesaurus Paradigm and Components

A data thesaurus is a special type of text database that organizes metadata much as controlled vocabulary thesauri organize indexing terms for bibliographic information systems [12]. But whereas bibliographic thesauri contain a fixed set of attributes for one type of entity (subject terms), data thesauri include different components concerning multiple types of metadata entities and their relationships.

A data thesaurus is composed of individual ***entries***, each of which represents an individual instance of a particular entity, attribute, or domain (set of comparable values). Each entry is in turn composed of discrete components that contain distinct types of metadata information. Exhibit 5, for example, shows a simplified hypothetical data thesaurus entry for a particular type of biotechnical entity -- namely a gene. Other entries might pertain to other types of entities (e.g., proteins), attributes (e.g., Location), or domain values (e.g., amino acid code values). In this example, components have been clustered into broad functional classes, as indicated at the left of the exhibit.

Function	Component	Component Value(s)		
Identity/Type	Identifier	2549685-7		
	Sequence Tag	...CAACGGTATAGGCTAACCTG ...		
	Type	gene		
Names/source(s)	Primary Name	SOD1		
	Prior Name (source)	IPO-A (mim)	SODS (hgml)	
	Used For	superoxide dismutase-1		
Internal links	Related Genes (RE)	SOD2		
	Alleles (NE)	SOD A*1	SOD A*2	
External links	MIM Reference(s)	14745		
	GenBank Sequence(s)	J02947	K00065	M13267
	HGML Locus Identifier	LM0147		
Book-keeping	Audit Trail: Date Modified	6/28/89	2/1/89	3/4/88
	Audit Trail: Modified By	jlmc	teh	bcd

Exhibit 5: Simplified Hypothetical Data Thesaurus Entry shows metadata components

Note that some components may be comprised of composite structures (e.g., Audit Trail contains two attributes). Some have multiple values (e.g., Alleles, Prior Name), and some may be constrained to specified domains (e.g., Sequence Tag contains only A,C,G, or T as values). Some components are pointers to other data thesaurus entries -- "RE" denotes Related Entries and "NE" denotes "Narrower Entries." As in object-oriented programming, broader class entities may contain generic information which all members of the class inherit, and classes may contain other classes recursively. Other components are identifiers (foreign keys) for data records in external databases.

Users and Uses

Data thesauri can be used in a variety of ways. Scientists may browse or query a thesaurus directly, or use it in conjunction with interface and data administration software to manage data and metadata integrity constraints for the DBMS; translate names into preferred terms or identifiers; "explode" a term to include broader, narrower, or related terms; convert between measurement units or coded value domains; and generate dynamic menus based on broader term/narrower term relationships.

The data thesaurus provides a bridge between diverse terminologies and databases. It does not impose a single nomenclature or classification standard. Instead, it provides a controlled environment within which multiple names, classification hierarchies, and value coding schemes can co-exist and evolve independently. It does not automatically resolve problems of terminology or data structure inconsistency, but it provides a framework for addressing such problems in a systematic way. Terminology managers can use the thesaurus to organize and document addition, deletion, and modification of entries over time (using attributes such as "Date Added," "Prior Term," etc.). Different individuals or groups may be responsible for disjoint sets of entities, attributes, or domains, so long as they do so within the over-all framework. Just as the metathesaurus currently being developed as part of the National Library of Medicine's Unified Medical Language System will unite controlled vocabularies from several bibliographic systems [13], the data thesaurus concept could be used to document and integrate work being done by various national and international standards efforts.

Implementation Issues

Experience with LBL's prototype material properties data thesaurus has demonstrated that implementation of this concept requires standard database capabilities plus special features for text data structures. LBL is currently exploring how different types of database software might meet these requirements. Commercial relational systems offer a mature, standard data model, query language (SQL), and program interfaces, but they are not well-suited for text and metadata [14]. Text management and retrieval systems offer much of the necessary text capabilities, but not a standard query language. Object-oriented database systems appear well-suited to many of our requirements, but few, if any have special features for text and the technology is untested.

SUMMARY CONCLUSIONS

The rapidly increasing amount, scope and diversity of biotechnical data made possible by computers has raised a host of terminology problems. Bioscientists urgently need better administrative and software mechanisms to develop and maintain controlled vocabularies and classification schemes necessary for effective data access, integration, and sharing. The data thesaurus is a new concept that addresses many of these issues. It builds on a familiar paradigm and previous metadata management tools to provide a systematic, extensible framework within which people and computer programs, standards organizations and individual research projects can:

- picture how different types of information relate to one another;
- reconcile nomenclature and classification systems from diverse, autonomous databases;
- support data and meta-data integrity constraints in a unified framework; and
- facilitate coordination between autonomous participating organizations.

ACKNOWLEDGEMENTS

This work was supported by the Director, Office of Energy Research, Health and Environmental Research, Human Genome Program, of the U.S. Department of Energy, under Contract DE-AC03-76SF00098.

REFERENCES

1 Cameron G (1989) "The GenBank Transaction Protocol -- The EMBL View," (EMBL Data Library).

2 Ingerman PZ (1982) "Make the Most of Your Check Digits," *Dr. Dobb's Journal* 66:36

3 Gumm HP (1985) "A New Class of Check-Digit Methods for Arbitrary Number Systems," *IEEE Transactions on Information Theory* IT-31:1 pp. 102-105.

4 McCarthy JL (1984) "Scientific Information = Data + Metadata," In Solomon H (Ed.), *Proceedings of the Workshop on Data Base Management, Monterey, CA* , pp. 79-124.

5 ANSI Acredited Standards Committee X3L8 (1990) "Coordination of Data Elements,"

6 Board of Regents NLM (1986) "Report of the Board of Regents, National Library of Medicine, NLM Long Range Plan Report of Panel 3: Obtaining Factual Information from Databases," (U.S. Department of Health and Human Services).

7 ISGN (1987)*Guidelines for Human Gene Nomenclature: An International System for Human Gene Nomenclature (ISGN, 1987)*, (S. Karger)

8 Fleck L. *Genesis and Development of a Scientific Fact* . (Chicago: U. Chicago).

9 Newton JJ (1987) "Guide on Data Entity Naming Conventions," *NBS Special Publication 500-149* (U.S. Department of Commerce, National Bureau of Standards).

10 McCarthy JL (1987) "Information Systems Design for Material Properties Data," In Glazman JS, Rumble JR Jr. (Ed.), *Proceedings of the Computerization and Networking of Materials Data Bases, Philadelphia* , pp. 135-150.

11 McCarthy JL (1988) "The Automated Data Thesaurus: A New Tool for Scientific Information," In CODATA (Ed.), *Proceedings of the 11th International CODATA Conference, Karlsruhe, Germany*

12 American National Standards Institute. Subcommittee 25 on Thesaurus Rules and Conventions. (1980) "American National Standard Guidelines for Thesaurus Structure, Construction, and Use," *ANSI Z39.19-1980, revision of ANSI Z39.19-1974*

13 Humphreys BL, Lindberg DAB (1989) "Building the Unified Medical Language System," *Proc Symp on Computer Applications in Medical Care, Washington, D.C.*

14 Lynch CA (1988) "Developments in Database Management System Technology and Their Impact on Information Retrieval," *Proc. ASIS Annual Meeting,* , pp. 190-194.

Biotechnology in the Encyclopedia of Chemical Technology – Standardizing the Vocabulary

Mary Howe-Grant

John Wiley & Sons, Inc., 605 Third Avenue

New York, New York 10158-0012

Abstract: The **ENCYCLOPEDIA OF CHEMICAL TECHNOLOGY** (**ECT**), a significant reference work for the chemical industry, serves as an instructional aid for scientists and engineers who are generally not specialists in the particular area where they are seeking information. The **ENCYCLOPEDIA** provides an introduction of a wide range of topics to a diverse readership. Because **ECT** is both a teaching tool and regarded as an authoritative source, the vocabulary within it must be standardized. Chemical and biotechnological terminology, nomenclature, symbols, units, etc. must be both consistent and accurate.

Biotechnology is not new to **ECT**. Coverage began with such entries as ENZYMES AND ENZYMOLOGY and FERMENTATION in the first edition and continues through the newest work, **ECT-4**, which is in preparation for the 1990's. Constructing **ECT-4** means choosing the 1180 article titles, identifying authors, and arranging for peer review, in addition to editing the material and keeping to schedule. This process, together with the requirements for standardizing the scientific vocabulary, will be discussed.

INTRODUCTION

The **ENCYCLOPEDIA OF CHEMICAL TECHNOLOGY** (**ECT**), a significant reference work for the chemical industry, serves as an instructional aid for scientists and engineers who are generally not specialists in the particular area where they are seeking information. The **ENCYCLOPEDIA** provides introductions on a wide range of technical topics to a scientifically diverse readership. Because **ECT** is regarded as a teaching tool as well as an authoritative source, the vocabulary within must be both accurate and consistent. As stated in the ECT Author's Guide, "an incidental purpose of this **ENCYCLOPEDIA** is to help develop clarity and, if possible, uniformity in chemical nomenclature."[1] The editors strive to present standardized chemical and biotechnological terminology, nomenclature, symbols, and units. Each edition of **ECT** has addressed this issue of nomenclature. Each edition speaks for itself in this endeavor and each is reflective of its times.

Biotechnology has been clearly and concisely defined by an international study group funded by the Organization for Economic Cooperation and Development as "the

K. Loening (Editor)
The Terminology of Biotechnology:
A Multidisciplinary Problem

application of scientific and engineering principles to the processing of materials by biological agents to provide goods and services."[2] Thus when the goods and services produced by biological means are chemical, biotechnology and chemical technology overlap. Chemists have used biological agents for the production of organic chemicals throughout history. Acetic acid, acetone, ethanol, and butanol have all been produced through commercial, economically-viable fermentation processes at one time or another in the latter half of this century. It should therefore be no surprise that biotechnology has been covered in **ECT** since the inception of this prestigious work.

ECT-1, THE FIRST EDITION

Inception: The **ENCYCLOPEDIA OF CHEMICAL TECHNOLOGY** was conceived in the early 1940's by the founders of Interscience Publishers, Inc.[3] The plan was to serve the burgeoning American chemical industry by providing technical information in the various fields of chemical technology. **ECT** would be a multi-volume work, a collection of state-of-the-art articles prepared by experts and presented in an alphabetized manner. Needed were a chemist and a chemical engineer to serve as editors of this heroic endeavor. The editorial role was to be similar to that in a peer-reviewed journal: manuscripts would undergo peer review with the editors determining the technical topics and the various authors. In 1944 Dr. Raymond Kirk, Professor of Chemistry at the Polytechnic Institute of Brooklyn, accepted the **ENCYCLOPEDIA** editorship. A co-editor was soon found in the person of his chemical engineering colleague Professor Donald Othmer and **ECT** was on the way to becoming a reality.

Interscience published the first of the fifteen planned volumes of **ECT** in 1947. Publication then continued at regular intervals through 1956. Rapid growth and technological advances in the chemical industry during the early 1950's precipitated two Supplement volumes, one in 1957, the other in 1960. Although Raymond Kirk's death in early 1957 preceeded the publication of the first Supplement, **ECT** is still widely known as the **KIRK**, or as the **KIRK-OTHMER ENCYCLOPEDIA.**

Standardized Vocabulary: Chemical nomenclature was a concern for **ECT-1**. A "Note on Nomenclature" appeared as Front Matter in Volume 1. Written by assistant editor Janet D. Scott, it states: "Two general purposes have been kept in mind: to reflect nomenclature usage in industry and to indicate other usages, particularly preferred usages. Therefore common or industrial names (as caustic soda instead of sodium hydroxide) have usually been employed in articles dealing with technology, but not necessarily in the more theoretical articles."[4] The preferred name is later defined as any of the International Union of Chemistry (IUC), **Chemical Abstracts**, or U.S. Pharmacopoeia designations. In addition to Volume 1's Front Matter, the article Nomenclature appears in Volume 9. This 12-page article is primarily concerned with inorganic and organic nomenclature. W. C. Fernelius of Penn State contributed the inorganic portion, Austin M. Patterson of Antioch College, the organic section. A one-paragraph section on biochemical nomenclature stresses the complexity of the subject.

Biotechnology: Table 1 lists the various **ECT-1** articles that could come under the modern heading of biotechnology. Included are substance articles if one of the principle means of manufacture involved biological agents. For example acetone is listed because about 20% of the acetone manufactured in the US in the early 1940's was produced by fermentation. Also listed are articles on biological agents useful to the chemical industry, such as enzymes and yeasts, and the process article Fermentation. Fermentation, a 58-page article published in Volume 9 in 1952, lists twenty organic chemicals, exclusive of pharmaceuticals, that are fermentation products.[5]

Table 1: Biotechnology Articles in **ECT-1** [6]

Acetic Acid
Acetone
Alcohol, industrial
Antibiotics
Beer and brewing
Butyl alcohols
Chloromycetin
Enzymes and enzymology
*Enzymes, industrial
Fermentation
Hormones
Lactic Acid
Penicillin
Pharmaceuticals
Polysaccharides
Propionic Acid
Proteins
Sterilization
Vinegar
Yeasts

*Article appeared in the First Supplement

ECT-2, THE SECOND EDITION

In 1961 John Wiley & Sons acquired Interscience Publishers, Inc. and plans for the **ENCYCLOPEDIA**'s second edition, **ECT-2**, were soon underway. This greatly expanded Wiley-Interscience edition reflected the concommittant growth in chemical technology during the 1960's. Anthony Standen, who had served as an assistant editor on **ECT-1**, and as editor after Raymond Kirk's death, became Executive Editor of **ECT-2**. Professor Othmer moved to the newly appointed Editorial Board where he was joined by Professors Herman F. Mark of Polytechnic and John J. McKetta, Jr. of the University of Texas. **ECT-2**'s twenty-two regular volumes came out at a steady rate from 1963 to 1970. These volumes were followed by a Supplement (1971) and an Index (1972).

Standardized Vocabulary: **ECT-2** contained both a fourteen-and-a-half page article Nomenclature and a piece on Units. The former was written by Kurt Loening of the **Chemical Abstracts Service.** It emphasized the International Union of Pure and Applied Chemistry's (IUPAC's) rules for inorganic and organic substances. The article's comments on biochemical nomenclature were similar to those in the previous edition. The article on units was written by M.L. McGlashan of the University, Exeter, U.K. for the Supplement in 1969. SI Units were introduced therein and the process of "metrification" in the U.K. mentioned. The author also expressed his hope that "metrification" would soon find its way to the US.

Biotechnology: The 1960's brought rapid growth and substantial changes to the chemical industry. Fermentation processes, although still important for the production of some organic chemicals, played a lesser role in chemical technology. The articles in **ECT-2** reflected that: Fermentation, a 58-page article in **ECT-1** was reduced to ten pages in this expanded edition. Table 2 lists **ECT-2**'s biotechnology articles. Although small quantities of acetone, 1-butanol, and ethanol continued to be produced by fermentation, the manufacturing processes of choice for these materials were now synthetic. Pharmaceuticals, however, began to play a larger role in the chemical industry.

Table 2: Biotechnology Articles in **ECT-2** [7]

Algal cultures
Antibiotics
Beer and brewing
Chloramphenicol
Enzymes, industrial
Fermentation
Hormones
Lactic Acid
Penicillins
Pharmaceuticals
Polypeptide antibiotics
Proteins
Sterilization
Vinegar
Yeasts

ECT-3, THE THIRD EDITION

Technological advances again dictated an up-dating of the material in **ECT** and a third edition of this highly respected reference work was planned. Acquisitions for **ECT-3** began in 1975 under the direction of the Executive Editor, Martin Grayson. Volume 1 appeared in 1978 and the twenty-six volume set, including Supplement and Index, was completed in 1984. The Editorial Board for **ECT-3** was composed of Professors Mark and Othmer from Polytechnic Institute of New York, Professor Charles G. Overberger from the University of Michigan, and Nobel laureate Professor Glenn T. Seaborg of the University of California at Berkeley.

Standardized Vocabulary: The emphasis on standardized vocabulary is evident throughout **ECT-3**. The Front Matter of each volume contains a 2-page Note on **Chemical Abstracts Registry Numbers** and Nomenclature in addition to a 13-page summary of Conversion Factors, Abbreviations, and Unit Symbols. Indeed, **Chemical Abstracts Services** names and **Registry Numbers** are given throughout as are SI Units. Because **ECT** is written primarily for the chemical industry, traditional units are simultaneously used. Additionally, common names for chemicals are often used in the **ENCYCLOPEDIA** after the **CAS** name has been identified. There is once again a Nomenclature article. It comprises some nineteen pages and is again written by Dr. Loening of **CAS**. Biochemical nomenclature has been expanded to a page and a half. The Units and conversion factors article is in the main body of **ECT-3**. It discusses SI Units almost exclusively.

Biotechnology: Table 3 lists the biotechnology articles that are present in **ECT-3**. The tenor of the list has changed from that of previous editions. Classes of biological agents, processes, and techniques specific to biomaterials have taken the place of the organic chemical substance articles. There are three separate pieces on enzymes and an article on microbial polysaccharides. Microbial transformations and genetic engineering, articles new to the **ENCYCLOPEDIA**, are representative of the new biotechnology. Fermentation is still an important process, the article even doubled in size from **ECT-2**, but the products are now primarily pharmaceuticals, alcoholic beverages, and, of course, vinegar.

Table 3: Biotechnology Articles in **ECT-3** [8]

Antibiotics: over 300 pages
Beer
Enzymes, immobilized
Enzymes, industrial
Enzymes, therapeutic
Fermentation
Genetic engineering
Hormones
Microbial polysaccharides
Microbial transformations
Pharmaceuticals
Pharmaceuticals, optically active
Proteins
Sterilization
Vinegar
Wine
Yeasts

ECT-4, THE FOURTH EDITION

The Plan of the **ENCYCLOPEDIA**

Work on the fourth edition of Wiley's **KIRK-OTHMER ENCYCLOPEDIA OF CHEMICAL TECHNOLOGY** began in July of 1988 under the direction of Executive Editor

Jacqueline Kroschwitz. **ECT-4** will be a state-of-the-art resource well-grounded in the tradition of earlier works but constructed to look forward into the 21st century. Publication is scheduled to begin in 1991, continuing through 1998.

This newest edition will comprise twenty-seven volumes. It will include a minimum of 845 main entries (1° words), 1185 peer-reviewed articles prepared by industrial and academic experts, and 25,000 pages excluding the Index. There will be over 500 pages devoted to articles associated with biotechnology and once again an emphasis will be placed upon standardizing the vocabulary of all substances and techniques presented in the **ENCYCLOPEDIA.**

There are two major parts to preparing a plan for the **ENCYCLOPEDIA.** The first is the compilation of the Word List: determining what material is to be covered and where in the **ENCYCLOPEDIA**, under what title, it should appear. The Word List is composed so as to reflect the state-of-the-art in chemical technology. The second part to preparing a plan involves author acquisitions: ascertaining who shall write the various articles.

<u>Compilation of the Word List:</u> Compilation of the **ECT-4** Word List began with the examination of article titles from **ECT-3**. Each **ECT-3** article was assigned to one or more of twenty-five editorial subject categories. These categories span the fields of coverage within the **ENCYCLOPEDIA.** They range from AGRICULTURAL CHEMICALS to WATER SUPPLY, PURIFICATION, AND REUSE. Two new categories developed for the purpose of **ECT-4** review are BIOTECHNOLOGY and COMPUTERS IN CHEMICAL TECHNOLOGY. Some categories (FATS AND WAXES) held as few as nine articles, others (INDUSTRIAL ORGANIC CHEMICALS) had over two hundred titles.

A preliminary, in-house editorial review of each subject category's list of article titles and page lengths followed. Articles such as Biosensors and Electronic Chemicals were added to the appropriate subject categories to reflect technological advances since the publication of the previous edition. Some **ECT-3** articles were considered for deletion, others retitled or reorganized. Genetic engineering, an entry in both the main body of **ECT-3** and in its Supplement, was separated into four articles, for example, each coming under the primary heading of Genetic engineering. The four new articles are designed to discuss 1) the techniques generally, 2) work with plants, 3) work with animals, and 4) microbial advances.

Each category's list of articles was then sent out to two or more experts in the particular subject field for examination. The comments of these outside reviewers were weighed against the total **ENCYCLOPEDIA** Word List. Subsequent changes were then made by the editors pursuant to the reviewers' comments and the entire plan. An additional in-house review was followed by volume assignation and the Word List was finalized.

Acquisition of articles: Author acquisitions for Volume 1 of **ECT-4**, tentatively Ablative Materials through Alkaloids, began in January of 1989. Authors are either industrial or academic experts who have state-of-the-art knowledge of their subjects. Potential authors are contacted up to fifteen months prior to the due date of the article. Completed manuscripts are always peer-reviewed and often technically edited to maintain the **ENCYCLOPEDIA**'s style and to standardize terminology.

Standardized vocabulary: Several steps have been taken to ensure the standardization of accurate terminology within this newest edition of **ECT**. First the Author's Guide contains no fewer than three pages of advice on nomenclature and five, including tables, on units.[1] A request for **CAS Registry Numbers** and approved names of chemicals is made of, and expected from, each author. **Registry Numbers** of many substances will be checked for accuracy with the **Chemical Abstracts Services**. Both SI and traditional units will again be used throughout the **ENCYCLOPEDIA**. Each volume will again have Front Matter emphasizing appropriate nomenclature and units. Articles such as Databases, Information Retrieval, Nomenclature, and Units and Conversion Factors will discuss preferred terminology. Then, last, but not least, the editors will strive to standardize the vocabulary as well as the style, throughout the work.

Biotechnology: Table 4 lists the biotechnology articles that will appear in **ECT-4**. Many of these titles are new to the **ENCYCLOPEDIA**. Indeed, this list consists primarily of articles dealing with the scientific and engineering principles of biotechnology such as Cell culture technology and Protein engineering. No longer included are many of the substance articles such as Antibiotics that represent the products of biotechnological efforts.

SUMMARY

Standardized vocabulary: Nomenclature and uniformity of terminology have always been a concern for **ECT**. Each edition has emphasized the standardization trends of its decade. **ECT-1** aimed to provide both practical and the current IUC, **CAS**, or US Pharmacoepia preferred names to substances. Later editions have continued in this vein, with **ECT-3** responding to the need of the chemical community to have **CAS Registry Numbers** as well. SI units have been found in the **ENCYCLOPEDIA** since **ECT-2**. **ECT-4** will again serve as an authoritative source with regard to scientific terminology whether it is in naming substances, in presenting preferred names, **CAS Registry Numbers**, or SI units, or in using the practical vocabulary of the 1990's. This task will be accomplished through the aid of authors, reviewers, and rigorous editing policies.

Table 4: Biotechnology Articles in **ECT-4** (Category 24)

* Aeration - Biotechnology
- Automated instrumentation - Clinical chemistry
- Automated instrumentation - Hematology
+ Biopolymers - Analytical techniques
+ Biopolymers - Survey
* Biosensors
* Biotechnology
* Cell culture technology
- Chromatography
- Enzyme applications - Industrial
- Enzyme applications - Therapeutic
- Enzyme processing
* Enzymes, inhibitors and antagonists
  Fermentation
+ Genetic engineering - Animals
+ Genetic engineering - Microbes
+ Genetic engineering - Plants
+ Genetic engineering - Procedures
* Immunoassay
  Immunotherapeutic agents
  Insulin and other antidiabetic drugs
  Medical diagnostic reagents
  Microbial polysaccharides
  Microbial transformations
- Nucleic acids
* Protein engineering
  Proteins
  Vaccine technology

Total pages for **ECT-4** Cat. 24 = 515

---

* Title and material are new to **ECT**.

- Title is new to **ECT**; the topic appeared elsewhere in **ECT-3**.

+ Primary title appeared in **ECT-3**; article titles have been changed as a result of reorganization.

Biotechnology: Biotechnology has waxed and waned in **ECT**. The use of biological agents was limited to fermentation in the early years of the US chemical industry. Because fermentation was the process of choice for manufacturing many organic chemicals in the 1940's and 1950's, these procedures appeared in **ECT-1**. Fermentation techniques took a backseat to a more sophisticated synthetic industry in the 1960's and 1970's. **ECT-2** and **ECT-3** articles reflect that. **ECT-4**, the edition of the 1990's, will feature many articles in biotechnology. Advances in molecular biology and microbial engineering have once again brought interest in the production of goods through biological agents to the fore. The edition will strive to present this information in terminology as standardized as the times will allow.

References:

1 Kroschwitz JI, Howe-Grant M (1989) Author's Guide **ENCYCLOPEDIA OF CHEMICAL TECHNOLOGY**, Fourth Edition, Wiley-Interscience, John Wiley & Sons Inc, New York

2 C&E News (1982) October 18:17

3 Mark HF (1988) J Pol Sci, Chem Ed 26:xv; Moore JH, WILEY, John Wiley & Sons Inc, New York

4 Scott JD in Kirk RE, Othmer DF (eds) (1947) **ENCYCLOPEDIA OF CHEMICAL TECHNOLOGY**, Interscience Publishing, Inc, New York 1:xv

5 Debecze GT in Kirk RE, Othmer DF (eds) (1951) **OP Cit** 6:317

6 Kirk RE, Othmer DF (eds) (1960) **ENCYCLOPEDIA OF CHEMICAL TECHNOLOGY**, Interscience Publishing, Inc, New York

7 Standen A (ed) (1972) **ENCYCLOPEDIA OF CHEMICAL TECHNOLOGY**, 2nd Edition, Wiley-Interscience, John Wiley & Sons, Inc, New York

8 Grayson M, Eckroth, D (eds) (1984) **ENCYCLOPEDIA OF CHEMICAL TECHNOLOGY**, 3rd Edition, Wiley-Interscience, John Wiley & Sons, Inc, New York

Nomenclature of Pharmaceuticals

Charles H. Barnstein, Ph.D. and William M. Heller, Ph.D.

The United States Pharmacopeial Convention, Inc.
12601 Twinbrook Parkway, Rockville, MD 20852

The United States Pharmacopeia (USP) and National Formulary (NF) are the official compendia of drug standards in the United States of America. Standards of the USP and NF also are recognized in the laws of numerous other nations.

The official compendia set forth standards of identity, strength, quality, purity, packaging, storage, and labeling of articles recognized, in USP, on the basis of therapeutic value and, in NF, as inactive ingredients suitable for use in dosage forms. Compendial standards are developed with a view to ensuring uniformity among drug products. Articles bearing the same name, whether from different lots of a manufacturer's product or from different commercial sources, purport to conform with the requirements of the official compendial monograph on that article. A drug not conforming to the official compendial standards, under the law, may be declared adulterated or misbranded. A single official title thus is a key factor in a drug standard.

NAMING DRUGS AND DRUG PRODUCTS (DOSAGE FORMS)

The federal Food, Drug, and Cosmetic Act includes certain provisions regarding drug names. The law refers to "established names" for drugs and defines such names to include the names given for drugs in the official compendia. The preparation of a drug standard includes the assignment of a name that is shorter and more useful than the unduly complex complete chemical name.

The United States Adopted Names (USAN) program is the specifically organized effort in the United States directed to producing simple and useful nonproprietary names for drugs. A name is adopted while a drug is still in its investigational

K. Loening (Editor)
The Terminology of Biotechnology:
A Multidisciplinary Problem

stage and before a variety of trivial names might otherwise find general usage.

The USP Committee of Revision has been given the authority to develop drug standards and name the compendial articles. The USP or NF titles are legally recognized as the designations for use in labeling the articles to which they apply. As a cosponsor of the USAN program, USP is committed to having the USP Committee of Revision use the USAN in developing the USP or NF standards for articles.

The USP Drug Nomenclature Committee (USP DNC) of the USP General Committee of Revision is charged with the responsibility for promoting uniformity and consistency among official titles for monographs on drugs in the United States Pharmacopeia (USP) and the National Formulary (NF) and the USP Dispensing Information (USP DI). The intent is to coin titles that are in harmony with convenience in physician prescribing and pharmacist dispensing and with accepted tenets of general usage and that are simple, useful, and clearly distinguishing and differentiating. The USP DNC reviews titles that predate the USAN program and titles for articles not considered by the USAN Council (e.g., inactive ingredients, medical devices, and diagnostic agents). One of the primary concerns of the USP DNC is the terminology of pharmaceutical dosage forms.

The USP DNC is advisory to the Executive Committees of USP's Drug Information Division and Drug Standards Division. To promote dialogue and stimulate exchanges of ideas and technology, USP

- o publishes its bimonthly journal Pharmacopeial Forum.
- o conducts USP Open Conferences on compendial issues.
- o participates in scientific and professional meetings.

These sources have proved to be valuable in assisting

- o USP DNC -- making recommendations.
- o Division Executive Committees -- making policy decisions.
- o USP Committee of Revision -- developing drug standards.

Readers of this paper who wish to participate are invited to examine the outlined subjects and to register preferences, comments, and suggestions with the authors.

DRUG NAMES IN MONOGRAPH TITLES

Drug Substance Monograph

The title names the drug form identified by the monograph as the acid, base, salt, ester, etc., for example:

- Codeine
- Codeine Phosphate
- Codeine Sulfate

Dosage-Form Monograph

The title might include the name of the drug form:

- Codeine Tablets
- Codeine Phosphate Tablets
- Codeine Sulfate Tablets

The title naming only the therapeutically active component (i.e., active moiety) is simpler and still useful. Thus, one title, instead of the above three, would be simply:

- Codeine Tablets

This ACTIVE MOIETY concept permits:

- one title (and monograph) instead of two or more.
- elimination of the inactive components of the molecule from the title.
- identifying the whole molecule on the label and/or in the labeling, not in the title.
- uniform labeling to identify inactive components, i.e., both inactive ions and excipients.
- simplified prescribing and dispensing.

Oral Solids -- Capsules; Tablets; Lozenges

Capsules and Tablets release active drug(s) promptly when swallowed. A descriptive term may be added to the title to indicate that the drug release is modified. The principle is that different articles should have different names:

- Tablet (prompt release, swallowed intact)
- Chewable Tablet (to be chewed before swallowing)
- Buccal Tablet (held in the mouth, where
- Sublingual Tablet the drug is released)
- Modified-release types --
  - Extended-release Tablet (sustained, repeat, or continuous slow release of drug)
  - Delayed-release Tablet (enteric-coated, so the tablet will pass through the stomach for release of the drug in the intestines)

Currently, only three terms are used in new USP titles for oral solids:

- Capsules (not further distinguished as hard capsules, soft capsules, or gelcaps) -- solid, liquid, or semisolid may be encapsulated in either hard or soft shell types.
- Tablets (not further distinguished as sugar-coated, film-coated, or caplets) -- various elongated or other geometric forms may be produced.
- Lozenges (not pastilles or troches).

Oral Solid is the single term for titles that, for simplicity and with appropriate labeling, could replace the first two of the above -- but the USP DNC hasn't favored this.

Oral Liquids -- Oral Solutions; Oral Suspensions

Traditional pharmaceutical classes, e.g., syrups (sweetened liquids) and elixirs (liquids containing alcohol), are not as commercially differentiated as in the past; for example, some products marketed as

- Syrups may be sugar-free and/or contain alcohol.
- Elixirs may be sweetened and/or contain no alcohol.

Currently, only two terms are used in new USP titles for oral liquids:

- Oral Solution (not Elixir or Syrup)
- Oral Suspension (not Magma or Syrup)

Oral Liquid is the single term for titles that, for simplicity and with appropriate labeling, could replace both of the above -- but the USP DNC hasn't favored this.

Preparations For Use In Eyes Or Ears

Eye preparations in USP are Ophthalmic articles:

- Ophthalmic Solution
- Ophthalmic Suspension
- Ophthalmic Ointment

Ear preparations in USP are Otic articles:

- Otic Solution
- Otic Suspension

The terms "Eye" and "Ear," as used by the British Pharmacopoeia (BP), instead of "Ophthalmic" and "Otic," are simple and they unambiguously identify the target organs --

Eye preparations in BP:

- Eye Drops
- Eye Lotion
- Eye Ointment

Ear preparations in BP:

- o Ear Drops

The term "drops" does not distinguish between solutions and suspensions and the importance of that distinction has not been determined among USP panelists.

Topical Preparations Applied To Skin Areas

Topical Liquids, many of which traditionally have been called Lotions, vary widely, embracing solutions and suspensions with or without alcohol, surfactants, suspending agents, etc.

More descriptive terminology is applied to new USP monographs:

- o Topical Solution (not Lotion)
- o Topical Suspension (not Lotion)

Topical Liquid is the single term for titles that, for simplicity and with appropriate labeling, could replace all of the above.

Topical Semisolids -- Two terms are used in USP titles:

- o Cream (oil in water emulsion type)
- o Ointment (water in oil emulsion or oleaginous type)

Former terms, gels and jellies, are not used here.

Topical Solid Carriers are of two types:

- o Plaster (drug acts locally, not penetrating skin)
- o Patch or Transdermal System (drug is transported across skin barrier and acts systemically)
  - o Patch is simple, short, and used widely in advertising and publicizing products.
  - o Transdermal System is more cumbersome, but descriptive, and finds use especially in professional literature.

Availability for generic usage is questionable -- a brand product is marketed with "Transdermal Therapeutic System" (TTS) as part of its name.

Topical Aerosol is the term in USP titles for solutions or suspensions delivered from pressurized containers with no intent to control dosage.

Preparations For Inhalation Therapy

Inhalation Aerosol is the term in USP titles for solutions or suspensions delivered into the mouth and on to the bronchial passages from pressurized containers designed to release measured doses. Other terms are considered to be more appropriate than "aerosol" for fine solids suspended in volatile propellants:

- Pressurized Inhaler
- Metered-dose Inhaler

Inhalant is the term in USP titles for vaporized drugs that are inhaled, usually via the nasal passages rather than by mouth.

Parenteral Preparations

Titles should include the word "Injection" or "Injectable" so that the intended use is clear:

- [DRUG] Injection -- a solution that is ready for injection.
- [DRUG] for Injection -- a solid drug, with or without added excipients, to which liquid vehicles are added to prepare a solution for injection.
- [DRUG] for Injection Concentrate -- a liquid preparation which, for safety, must be diluted with a solvent prior to injection.
- Injectable [DRUG] Suspension -- a suspension that is ready for injection.

- o Injectable [DRUG] for Suspension -- a solid drug, with or without added excipients, to which liquid vehicles are added to prepare a suspension for injection.

Combinations Of Two Or More Drugs

Titles that combine the drug names may be long:

- o Sulfamethoxazole and Trimethoprim Tablets
- o Triamterine and Hydrochlorothiazide Capsules

Shorter, useful titles could eliminate the necessity for writing out the individual drug names. Such names that could represent the cited combinations are:

- o Co-trimoxazole Tablets
- o Co-triamterzide Tablets

The "Co-" prefix is used to identify the article as a combination product, and the remainder of the name uses stems from the names of components. The lengthy combination of full drug names is still used in the USP as the official monograph title, but the shorter "Co-" name is listed as a Pharmacy Equivalent Name (PEN) to facilitate prescribing and dispensing practices. Alternatively, as in the BP, the "Co-" name could be adopted as the official title.

These nomenclature initiatives are under continuing review by the USP DNC. Evidence of the implementation of some of the principles of nomenclature discussed in this presentation are found in the current volume of USP-NF.[1] Other titles that are proposed by the USP Committee of Revision are published, for review and comment, in Pharmacopeial Forum, USP's journal of drug standards development and official compendia revision.

Reference

1 The United States Pharmacopeia (USP XXII) and The National Formulary (NF XVII), United States Pharmacopeial Convention, Inc., Rockville, MD, USA (1990).

Author Index

Merritt, J. E., Chemical Abstracts Service, P.O. Box 3012, Columbus, Ohio 43210, USA (pp. 25-40)

Moo-Young, M., Department of Chemical Engineering, University of Waterloo, Waterloo, Ontario, Canada N2L 3G1 (pp. 165-170)

Myers, M.W., Division of AIDS, National Institute of Allergy and Infectious Diseases, National Institutes of Health, Bethesda, Maryland 20892, USA (pp. 61-70)

Schiffman, D.O, American Medical Association, 535 North Dearborn Street, Chicago, IL 60610, USA (pp. 109-114)

Secrist III, J. A., Kettering-Meyer Laboratory, Southern Research Institute, P.O. Box 55305, Birmingham, AL 35255-5305, USA (pp. 17-24)

Smith, D.L., Syntex Research, 3401 Hillview Avenue, Palo Alto, CA 94304, USA (pp. 93-108)

Smith, W.L., Department of Biochemistry, Michigan State University, East Lansing, MI, USA (pp. 93-108)

Sonneveld, H., TOPTERM C.V., European Division, W.G. Plein 278, 1054 SE Amsterdam, The Netherlands (pp. 171-186)

Tada, T., Department of Immunology, Faculty of Medicine, University of Tokyo, Hongo 7-3-1, Bunkyo-ku, Tokyo 113, Japan (pp. 125-134)

Tsugita, A., Japan International Protein Information Database (JIPID), Research Institute for Biosciences, Science University of Tokyo, 278 Noda, Japan (pp. 9-16)

Webb, E. C., F.R.A.C.I., Department of Chemistry and Biochemistry, James Cook University of North Queensland, Townsville, Queensland 4811, Australia (pp. 51-60)

Willis, A.L., Syntex Research, 3401 Hillview Avenue, Palo Alto, CA 94304, USA (pp. 93108)

Yamamoto, I., Tokyo University of Agriculture, Setagaya-ku, Tokyo 156, Japan (pp. 147-154)

Zorn, M.D., Human Genome Center, Lawrence Berkeley Laboratory, Berkeley, CA 94720, USA (pp. 85-92)

GPSR Compliance

The European Union's (EU) General Product Safety Regulation (GPSR) is a set of rules that requires consumer products to be safe and our obligations to ensure this.

If you have any concerns about our products, you can contact us on ProductSafety@springernature.com

In case Publisher is established outside the EU, the EU authorized representative is:

Springer Nature Customer Service Center GmbH
Europaplatz 3
69115 Heidelberg, Germany

Batch number: 10704771

Printed by Printforce, the Netherlands